CONTENTS

CHAPTER 3
Basic Electronics of Measurement

CHAPTER 4
Oxygen Measurement

CHAPTER 5
Carbon Dioxide Measurement

CHAPTER 6
pH

CHAPTER 11
Flow and Velocity Measurement

CHAPTER 12
Force, Displacement, and Pressure Measurement

CHAPTER 13
Measurement of Ions and Solution Properties

CHAPTER 14
Radioisotope Techniques

CHAPTER 15
Tracer Methods and Compartmental Analysis

Appendices

PREFACE

Natural science, we are taught, comprises a body of specific knowledge about the universe, described by general laws arrived at inductively from the particulars. Insofar as natural science is quantitative, the particulars must consist of measurements. Much of our lives as scientists, then, is concerned with the making of these measurements, and the validity of our generalizations, as well as the accuracy of our insight, depends on our ability to make accurate and appropriate measurements. Much of our formal education, however, is concerned with the theoretical side of the natural sciences; this is the stuff from which textbooks and lectures are made. It is only in the increasingly rare laboratory courses that we are given tutoring in the art of observing and measuring.

This book is about making measurements: not only *how* any particular method works, but *why* it works. The stimulus for writing it has come partly from experience with students in physiology courses and partly from observing the trends of change in university curricula. As scientific research has become more sophisticated, complex, and technologically oriented, the budget for undergraduate laboratory courses has increased at a much slower pace, if at all. Thus it is not rare to have first-year graduate students in the laboratory who have never actually made a pH measurement or worked with an oxygen electrode. One goal, then, has been to provide some compact reference for the commoner methods in use in physiology as an entry point for the student without such exposure. A second, and equally important, goal has been to provide the same entry point for working scientists who may simply not have needed some particular technique in the past. Which of us could have predicted ten years ago what techniques we would need, and which measurements would seem important?

What I have tried to emphasize is the basic principles underlying each of the methods and pieces of apparatus discussed. Manuals for instruments and instruction booklets for various methods are of variable lucidity on principles. Anyone can learn to "twiddle the knobs," provided they first have a basic understanding of the principles of the gadget. Proprietary manuals and methods booklets are rarely forthcoming on the pitfalls and limitations of any measuring technique, so this is another area I have tried to stress. References are given at the end of each chapter—not an exhaustive list but selected to direct the reader to more detailed discussions of each point if needed.

A reasonably broad coverage of measurement types has been included, though not all of the many measurements made by physiologists could be included. The introductory chapters deal with some fundamental concepts of measurement itself: basic gas and solution concepts, and a chapter on electronics relevant to measurement methods. The respiratory gases, oxygen and carbon dioxide, are sufficiently important to warrant their own chapters, as well as some further discussion in a chapter on blood gas measurement. The concept of pH as well as its measurement and control (buffering) are discussed, as are problems of controlling the gaseous environment. Basic principles of flow, velocity, force, displacement, and pressure are described along with common methods for their measurement. The final chapters deal with ions and solutions, radioisotope concepts and techniques, and tracer kinetics. Most of these topics apply to a broad spectrum of physiological studies, whether of membrane transport in single cells or the respiration of an exercising athlete. Although electrophysiological methods are not specifically treated, the general concepts of electrical transduction, amplification, and recording are described.

A final concern in such an undertaking is obsolescence: new devices appear almost daily, and the phrase "state of the art" is getting worn. New principles, however, evolve rather slowly, and a good grasp of them is perhaps the working scientist's best tool for staying abreast.

That this book was written is probably due in equal part to the enthusiastic response of my students to what might be called an introduction to the "tools of the trade," and to constant encouragement from my wife, Sharon.

James N. Cameron

CHAPTER
1

MAKING MEASUREMENTS

1.1. INTRODUCTION

In our theoretical upbringing as scientists, we think wholly in terms of an ideal world, one in which *parameters* have absolute values which may be determined with absolute precision. Such absolute terms never exist in the real experimental world, however, where we must deal with measured *variables*. To take a simple example, temperature is a fundamental parameter in many physical, chemical, and physiological relationships. We define temperature in an absolute sense as an amount of heat energy embodied in the molecular motion of various precisely defined systems, but temperature cannot really be sensed in a practical way by referring to such systems. In order to measure temperature, we employ a *thermometer*, literally a meter, or measuring instrument for temperature. The commonest sort is a mercury thermometer, which consists of a bulb filled with mercury, a glass column attached to it into which the mercury may rise as it is heated, and graduations marked on the glass. As we read the temperature by estimating the alignment of the mercury meniscus on the scale of graduations, we are far removed from the pure, philosophically absolute concept of temperature. Consider the possible sources of error and the imprecision of this simple measurement.

The principle of operation of this mercury thermometer is that by measuring the thermal expansion of mercury, an accurate estimate of the temperature will be obtained. In a typical mercury thermometer, the scale might be graduated, say, from -10 to $+100°C$ in $0.5°$ divisions. With care the scale reading may be estimated to within $0.1°$. In making the reading, however,

1

we have introduced the first source of error, since we may not estimate the tenths properly; our eyes may be misaligned, introducing parallax error; or we may simply make a mistake in recording the scale reading. If only the bulb is immersed in a fluid appreciably different from the air surrounding the shaft, another error is introduced, since the thermal expansion of the mercury in the glass column is not the same as that in the immersed bulb. If the bulb is not allowed to come to complete thermal equilibrium with the medium to be measured, another error will result from this source.

Finally, when the thermometer was manufactured, it had to be *calibrated*. This means that the temperature registered by the thermometer had to be compared to that measured by some other instrument, presumably of known accuracy. The subject of primary and secondary temperature standards is actually quite complex; various standards have been promulgated by the National Bureau of Standards and by international bodies, and a number of books have been written on the subject of temperature measurement. The result is that a manufacturer usually guarantees that the reading from a mercury thermometer will fall within specified error limits of the true temperature. For ordinary thermometers the range is usually $\pm 1°$. For thermometers of this type, it is clearly of little value to try to read them to the nearest $0.1°C$, unless one is interested in temperature differences, and not in the absolute value of the temperature. Certified thermometers may be purchased, at substantially greater cost, that provide accuracy of $0.1°C$ or better, but in using them the other errors mentioned above must be taken into account. For measurement of very small temperature differences, special differential thermometers (the "Beckman" type) are available which reliably register differences of only about $0.002°C$.

The use of a common mercury thermometer, then, entails possible errors from many sources: inadequate equilibrium, parallax in reading, non-uniformity of column and bulb, heterogeneity of the measured solution, and calibration tolerances. Whether these errors in sum are acceptable for the intended purpose must be evaluated constantly by the user, and if not, appropriate steps taken to reduce the magnitude of the error from each source. Complete elimination of experimental (measurement) error is never possible.

1.2. THE DISTRIBUTION OF ERROR

1.2.1. Systematic Errors

A systematic error, or *bias*, is by definition an error of the same magnitude and direction introduced equally into each of a series of measurements. To continue with the example used above, a calibration error in a mercury thermometer would represent a systematic error, or bias, when the thermometer was used for a series of measurements. Since an error of this kind would change the measured temperature by the same amount for each reading, it would have no effect on the measurement of temperature differences so long as the same thermometer was used throughout the series of measurements. It can also be corrected for very easily if discovered later. Other examples of systematic errors are an error in making up an acid for titration or dirt on a balance pan. In both cases, the error introduced into the measurements is the same for each of a series; difference measurements are not affected, and a systematic correction may be applied later to eliminate the error.

Not all systematic errors, however, are constant or so easily dealt with. To continue with the temperature example, let us suppose that an impatient investigator is using a mercury thermometer to measure a series of solutions, each substantially different from room temperature. This investigator tends to take the readings a little too quickly, before the thermometer has completely equilibrated with the solutions. A systematic bias will be introduced into this series of measurements, but it will not be of constant magnitude. A thermometer follows an exponential curve of approach to equilibrium, so the magnitude of the error will depend on the difference between the solution temperature and the room temperature, and the direction of the error will depend on whether the measured solution is hotter or colder than room temperature. This sort of error is very difficult, if not impossible, to correct for when discovered later, and affects measurements of differences as well as absolute measurements.

Still more subtle bias may be introduced by preconceived notions in the experimenter's mind of how a measured variable will turn out. These unconscious errors, which are the motivation for "blind" experiments, may take a great variety of forms. All of us must be constantly aware of the possibility of this "expected result" bias and must take steps to avoid it. Standardization of measurement methods is often a great help. To use the thermometer example once more, one might determine that it takes at least 1.5 min for the thermometer to come to complete equilibrium and then use a timer to ensure that all measurements are made after at least 1.5 min.

1.2.2. Random Errors

By definition a random error is one that occurs without pattern or predictability, although the science of statistics is concerned in large part with determining the pattern and predictability of numbers or series of random errors. In the thermometer example, random errors might result from reading the thermometer at different angles at different times, from allowing equilibration to proceed for irregular times, or from fluctuations in room temperature which would affect the glass column expansion. In the case of temperature measurement these errors might not be particularly large or serious, but in many other experimental measurements they become important. It is also the ratio of the magnitude of errors to the magnitude of the measured variable or differences in the variable (the "signal-to-noise ratio") that becomes important. For example, if one is trying to measure differences of $0.1°$, the random errors will very likely be quite significant.

Consistency of technique is undoubtedly the single most important way of reducing random errors, whether in the measurement of temperature, volumetric and gravimetric measurements, or any other experimental procedure. Even apparently insignificant details may contribute to experimental precision. For the temperature example, a good procedure for maximum precision in measuring differences might be to always set the thermometer in the same place when not in use, to insert it into the measured solution in exactly the same fashion each time, to allow it to come to equilibrium for exactly the same length of time in each solution, to insert the thermometer to exactly the same depth each time, and to make the reading facing the thermometer at the same angle and height each time. Above all, an evaluation of error magnitude should be carried out for critical measurements, for example by repeated alternate measurements of the same two solutions, and comparison of the replicate measurements.

1.3. PIPETTING: A BASIC LABORATORY TECHNIQUE

1.3.1. Introduction

Since so many procedures in physiology and physiological chemistry depend upon the accurate measurement of volumes of a great variety of liquids, the subject of pipetting is sufficiently important and complex to warrant separate treatment. Translated literally from the original French, a *pipette* is simply a small glass tube, but a check of a catalog from any major scientific supply house will reveal that there are hundreds of kinds of pipettes manufactured, many of which are far removed from a simple glass tube. To mention a few types, the pipettes may be graduated or volumetric; may have a bulb or not; may have finely drawn tips or coarse ones; may be made of special glasses; may be intended for special fluids or applications; may be of glass, plastic, metal, or combinations of these materials; may be permanent or disposable; and may be purchased in a variety of accuracy grades. Knowledge of the proper choice of pipette and the technique for its use may often be absolutely critical to the success of a measurement or experiment.

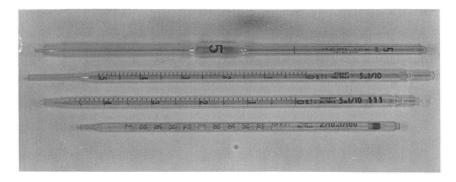

Fig. 1.1 Some common types of glass pipettes, including (from the top) a 5 ml volumetric type, a 5 ml graduated TC type, a 5 ml blow-out type, and a 0.2 ml capillary glass graduated pipette.

1.3.2. Glass Pipettes

Volumetric Pipettes. A volumetric pipette, such as that shown in Fig. 1.1, usually consists of a straight glass tube with a bulb blown in the middle, a drawn tip, and a volume mark above the bulb which may or may not be highlighted by a colored strip on the glass. The most common type of volumetric pipette is marked with its volume, the letters "TD," and a temperature. The "TD" designation ("To Deliver") means that when the fluid is drawn so that the meniscus lines up exactly with the volume mark and is then allowed to run out (with its tip touching a wetted surface) but not blown out completely, the volume marked on the upper end will be delivered when the solution is at the specified temperature and has the viscosity and surface tension of pure water. This seems straightforward, but errors may be introduced at a number of steps.

Volumetric pipettes are supplied in different grades, depending upon the accuracy of calibration guaranteed by the manufacturer. As an example, a Corning #7100-10 pipette, nominally a 10 ml volumetric type, has a stated calibration accuracy of ± 0.02 ml, which means that there may be as much as 0.2% error in the volume delivered under otherwise ideal conditions. The repeated use of a particular pipette, then, will introduce a systematic error of up to this amount, and the use of several similar pipettes will introduce a random error of up to this amount. If greater accuracy is required, certain manufacturers offer volumetric pipettes with individual certificates of calibration, but other errors involved in their use are usually at least this large, and the extra expense is rarely justified. Pipettes of lesser or unknown accuracy grades may be individually calibrated by filling them with mercury and weighing it, taking the temperature and density of the mercury into accurate account. The volume remaining in the tip after delivery is sometimes difficult to adjust for, however, so the method is probably more suitable for comparing pipettes with each other than for obtaining an absolute calibration.

The technique of filling and delivery can make a substantial difference in the accuracy achieved with volumetric pipettes. If the tip of the pipette is not touching a wetted surface, the tip will retain a greater variable volume of liquid, depending upon the size of the drop left hanging on the tip when delivery is complete. The amount contained in the tip when delivery is complete will also vary with the surface tension of the liquid— pipetting of ethanol or acetone, for example, should be done with a different type of pipette, as should the pipetting of viscous fluids. Since most fluids will wet the inside of the pipette, rapid emptying may leave more fluid on the inside walls than slow emptying. With liquids of viscosities greater than that of pure water, this becomes extremely important, and good technique dictates a slow and reproducible rate of emptying. All of these technique points become more important as the volume of the pipette decreases, since the same absolute volume error will constitute a progressively greater percentage error.

Certain kinds of volumetric pipettes may be marked "TC" to signify that the pipette is made "To Contain" the nominal volume. This means that the entire pipette volume should be delivered by blowing out the portion remaining in the tip after drainage. These pipettes are also called "blow-out" types, and are most often used for the measurement of fluid of higher or lower viscosity or surface tension than pure water. Most glass capillary micro-pipettes are made to be used in this fashion, as are a number of special-purpose pipettes (see Fig. 1.1).

Graduated Pipettes. Pipettes manufactured with a series of graduated volume marks (Fig. 1.1) are useful, of course, when irregular volumes must be delivered. It should be noted, however, that these pipettes usually have somewhat greater calibration tolerances than volumetric pipettes. For example, a Corning #7063 Mohr type measuring pipette with a nominal volume of 10 ml has a stated manufacturing tolerance of 0.06 ml, three times the possible error associated with the 10 ml volumetric type cited above.

Measuring, or graduated, pipettes are also supplied in a variety of types for general or special purposes and may have either the TC or TD designations, depending upon whether they are designed to deliver between marks or to deliver their complete volume by blowing out. Besides the letter markings, most manufacturers also code their pipettes for volume and type with colored and/or etched rings at the upper end.

Glass Micro-Pipettes. For the measurement of very small volumes, say less than 0.1 ml, with high accuracy, several different types of glass micro-

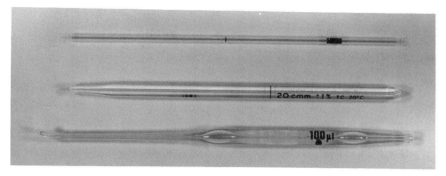

Fig. 1.2 Glass micro-pipettes, including (from the top) a disposable 100 μL capillary type, a 20 cmm (0.020 ml) hemoglobin pipette, and a 100 μL volumetric type.

pipettes have been developed. The re-usable types (Fig. 1.2a,b) may or may not have an expanded volume bulb and may or may not be used with a rubber tubing mouthpiece, but the general considerations for their use are similar to those for other volumetric pipettes. Most of them are of the TC type and must be blown out, since the effects of surface tension in very small bore tubes are large and difficult to control. In such small bore tubes, the amount remaining on the inside barrel wall may easily comprise a significant error, so the user must know whether the volume calibration assumes that the pipette will be rinsed, thus delivering its entire contents including the amount sticking to the walls. One must also be cautious in filling these micro-pipettes: the usual procedure with macro pipettes is to fill them to some (variable) point above the volume mark, and then allow the fluid to fall until it just reaches the mark. If this procedure is followed with a micro-pipette, particularly with a viscous solution (see below), a significant error may result upon subsequent rinsing, since some appreciable extra volume may remain stuck to the barrel walls above the volume mark.

Some of the disposable types of glass micro-pipettes (Fig. 1.2) circumvent this filling problem. They are designed to be completely filled, and so long as there is no extra material clinging to the outside of the tube, the volume delivered after rinsing should be unaffected by the filling procedure. These disposable capillaries are usually designed to be rinsed (TC) in order to deliver their full volume.

1.3.3. Pipetting Devices

Plastic Tip Types.   In recent years, pipettors constructed with a spring-loaded plunger and a disposable plastic tip which acts as the volume container have become quite popular (Fig. 1.3). These have a number of advantages over glass pipettes: 1) they do not have to be cleaned between uses, 2) they are fast, 3) the positive-displacement plunger arrangement ensures that the

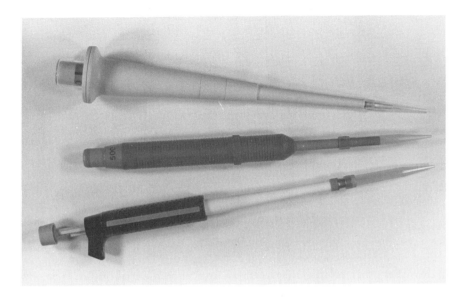

Fig. 1.3   Some common pipetting devices. Top: A 40 μL Eppendorf with an automatic tip ejector; middle: a 500 μL plunger type; bottom: a three-volume selectable type.

volume delivered does not depend upon the viscosity and/or surface tension of the liquid, and 4) The same plunger is always used for filling, so the reproducibility tends to be quite good. Unfortunately there is a natural tendency to regard automatic things as foolproof and error-proof, and this is certainly not the case with these pipettors. As with nearly any other technique, consistency and attention to detail are required for the best results, and sloppy technique can lead to amazingly large errors even with these apparently foolproof devices.

   To begin with, the volume taken up into the plastic tip with each excursion of the plunger may vary for several reasons. If the tip is not firmly seated each time, it may leak air, reducing the fluid volume. The plungers require periodic maintenance, lubrication and replacement of O-rings, and if these leak, either during fluid filling or before delivery, the contained volume is reduced. Erratic operation of the plunger may cause variable errors; for example, if the plunger is operated (released) rapidly, a large negative pressure may be generated, causing a small leak, but slow operation of the plunger on the next sample may not lead to the leak. Good technique requires even, fairly slow operation of the filling plunger and equally even, slow ejection. It is actually not completely obvious why this should make a difference, but pipetting and weighing of 10 or 20 replicates using first an erratic technique and then a steady, consistent one should convince anyone that it does make a difference.

A final point of technique pertains to the emptying of such pipetting devices. When pipetting into another liquid, the diluent should be drawn into the plastic tip and then ejected. This rinse cycle should be repeated several times to ensure complete delivery of any material sticking to the tip. Although the tips are usually non-wettable with pure water, many physiological fluids contain surfactant materials that do cause wetting, increasing the importance of rinsing. When pipetting onto a dry surface, a consistent contact angle should be used, and the tip moved as the delivery proceeds to maintain a constant shallow penetration of the tip into the ejected drop. This technique reduces the loss of considerable material along the outside surfaces of the plastic tip and helps keep the small loss consistent. Since rinsing is not possible, slow ejection will help minimize the volume retained on the inside of the plastic tip. For viscous liquids, it is probably not possible to deliver very accurately onto a dry surface with these devices, and this should be checked by weighing some aliquots of either the sample or a similar solution.

Syringes. For many applications, a syringe is superior to a pipette for volume measurement of liquids. This may due to physical requirements (delivery or filling via a small aperture), but is often due to the characteristics of the sample and/or the volume delivery requirements. Liquids can be transferred to and from a syringe without contact with the atmosphere, for example, and highly viscous materials may be delivered accurately by the positive barrel displacement. Fractional microliter volumes may be delivered with many small-volume syringes, and a variety of different materials for both barrel and needle accommodate special needs (Fig. 1.4). One particular feature of syringes that needs comment is the "dead space," that volume contained between the end of the plunger and the tip of the needle that is not emptied by full excursion of the plunger. If exposure to air is to be avoided, this space must be filled with something other than air, which may require a dilution correction to the final delivered sample. Special syringes with no dead space are available (Fig. 1.4), and are especially useful for dispensing small quantities of valuable materials such as hormone solutions or radioisotopes.

1.3.4. Additional Notes on Pipetting Technique

The importance of a steady and consistent technique cannot be over-emphasized—pipetting is probably done better by machines than by people, so the development of a machine-like technique should be the goal. Pipettes should be filled in a slow and steady manner, and should be emptied in the same way. The more viscous a solution, the slower should be the delivery time. With blood serum, for example, the emptying time for a Van Slyke serum pipette should be around 30 sec for 50 μL (Van Slyke & Plazin, 1961).

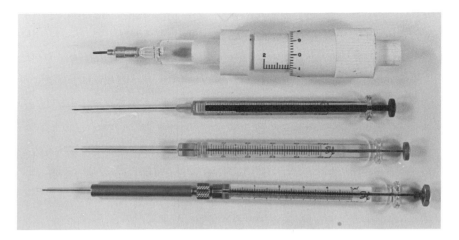

Fig. 1.4 Syringes for pipetting. From the top: A micrometer syringe for delivering 0–2000 μL in 0.002 ml increments; a 500 μL "gas-tight" syringe; a 50 μL steel-plunger syringe; a 5 μL zero dead space syringe.

For TD types of pipettes, use with viscous solutions should be accompanied by a check of the volume calibration with a solution of similar viscosity, since the amount remaining in the tip after delivery will vary with the viscosity. Very viscous solutions and organic solvents with low surface tension should probably be dispensed in TC type pipettes when high accuracy is required.

Even the method of blotting the outside of the pipette tip is important. The tip should be wiped by moving toward the tip along the shaft, but without making flat contact with the tip. Blotting perpendicular to the tip will cause the loss of some fluid out of the pipette by capillary action. It is often best to wipe the shaft and tip before adjusting the volume to the mark, then to carefully wipe any adhering fluid from the tip, and finally to re-check the volume mark to make sure that there has been no loss prior to delivery.

1.4. ACKNOWLEDGING ERROR WHEN DESIGNING EXPERIMENTS

There is a large body of statistical literature that deals with the subject of error and experimental design, but most of it is concerned with confidence limits, numerical processing of data, etc. There is an additional aspect to the subject, however, which has to do with a grasp of the likely error magnitude of various experimental procedures and with ways to design experiments so that the procedures with the highest error are least critical to the outcome. To illustrate, we will consider the problem of measuring salt fluxes in an aquatic animal.

For the example, assume we have a 50 g animal whose rate of Na^+ influx is 400 mEquiv kg^{-1} hr^{-1} and whose rate of Na^+ efflux is 450 mEquiv kg^{-1} hr^{-1}, resulting in a net loss of 50 mEquiv kg^{-1} hr^{-1}, or a total loss of 2.5 mEquiv hr^{-1}. If we know we can measure changes in $[Na^+]$ in the water with an accuracy of 2% and want to design an experimental aquarium to hold the animal, our analytical accuracy will determine the maximum volume, given the time intervals for which loss measurements are desired. For intervals of 15 min and a fresh water $[Na^+]$ of 1.0 mEquiv L^{-1}, the Na^+ loss in 15 min will be 0.63 mEquiv. This is 2% of the Na^+ content of about 31 L of water, but of course we would like the changes in the measured variable, $[Na^+]$ in this case, to be several times larger than the "background" analytical error. Using signal-to-noise ratios of 5:1 or 10:1 would dictate an experimental system volume of 6 to 3 L. If even a rough guess of the magnitude of experimental changes is known in advance, "thumbnail" calculations of this type, coupled with the likely percentage error of analytical procedures to be used, can prevent much loss of time. It is much faster to do these calculations than to discover at the end of an experiment that the initial design was such that the errors swamp the changes to be measured.

Closely related to the design of an experiment is the recognition of which steps are critical in a particular procedure and which are not. The conservative approach, of course, is to take maximum care with every step, but since this often is much more time-consuming and expensive, it is useful to know at which steps errors have the greatest effect on the final result. A good example is the use of a titration method to measure the net acid or base output of an animal in sea water. Under normal circumstances, the net output is usually within ± 200 μEquiv kg^{-1} hr^{-1} of zero; so for a 1 kg animal in a 10 L system, this means changes of less than 20 μM L^{-1} hr^{-1}. The normal titratable alkalinity value for seawater is about 2200 μEquiv L^{-1}, so the changes are no more than 1% per hour. The titration has several possible error sources, including errors in assessing the normality of the acid used for the titration, errors in end point, etc. Since the final value for the animal, however, depends upon differences between one titration value and the next, some simple calculations will show that the measurement (by pipette) of the aliquot to be titrated is the most critical step in the whole procedure. An error of $+0.5\%$ in the initial aliquot volume and of -0.2% in the final aliquot volume will result in an apparent -0.7% change in the titration value, with no change due to the animal. A change of -0.7% of 2200 is 15.4 μEquiv L^{-1} and, in a 10 L system, 154 μEquiv. This error is as large as what we are trying to measure, yet the pipette errors assumed were not unreasonable.

With maximum care, the volumetric error in aliquot measurement may be kept to 0.05% or less, but even this much error will translate into an error of 10 to 20 μEquiv acid or base excreted. Clearly the volumetric error associated with this single step, pipetting the sample aliquot for titration, is the major source of error in the entire experimental procedure, and other errors will be relatively insignificant. Again, some thumbnail calculations of this sort serve to highlight the critical steps, and will focus attention on steps in which special procedures and particular care are required. To conclude the example, if the titration is performed with an indicator, it would clearly be a waste of time and effort to take special precautions for measuring and dispensing it, since indicators are nearly always added in excess, and dilution has virtually no effect on the final result.

LITERATURE CITED

Van Slyke, D. D. & J. Plazin. 1961. Micromanometric analyses. Williams & Wilkins, Baltimore. 89 pp.

BASIC GAS AND SOLUTION CONCEPTS

2.1. INTRODUCTION

The basic life processes take place in aqueous solution and involve the exchange of respiratory gases, so a study of physiological function can hardly proceed without a thorough understanding of the physical properties of gases and liquids. Equally, the great variety of measurements involving gases, liquids, and solutions is based upon one or more physical or chemical attributes of the materials of interest. At one time or another, nearly every distinctive property of the respiratory gases, for example, has given rise to a method for measurement.

The discovery of these principles makes interesting reading in the history of science, with most of the great advances coming between 1650 and 1850. These included the discovery of the relationship between pressure and volume by Boyle (ca. 1650), the discovery of oxygen by Priestley in 1774, the elaboration of the gas laws by Charles in 1787 and Gay-Lussac in 1802, and the refinements provided by kinetic theory in the mid-19th century. The application of new technology to this fundamental body of knowledge has provided the modern physiologist with a great variety of powerful measurement tools.

2.2. GASES

Both gases and liquids are *fluids*, a fluid being defined as a state of matter in which the molecules are relatively free to change their position with respect to each other in a continuous fashion. The ease with which positional change can occur varies widely from one fluid to another, depending on the nature of the substance and the kinetic energy of the molecules.

A *gas* is a fluid in which the molecules are practically unrestricted by cohesive forces, and so a gas has no definite shape and no intrinsic volume. Both shape and volume are imparted by the container of the gas, and the *ideal gas law*, a combination of Boyle's and Charles' laws, states the relationship between pressure, volume, and temperature:

$$PV = nRT \qquad \text{(Eq. 2.1)}$$

where: P is pressure, V volume, n the number of moles of gas, R the gas constant, and T Temperature, Kelvin scale. The value for R varies depending on the units employed for pressure and volume; if volume is expressed in cm^3 and pressure in mm Hg (see 2.3), the value for R is 6240. The ideal molar volume for a gas is 22.414 L, but the gases of physiological interest have significantly non-ideal behavior. The molar volumes for some commonly measured gases are given in Table 2.1.

2.2.1. Pressure

Pressure is defined (and sensed) as the force exerted on the walls of the container of a gas. The pressure depends on the number of molecules contained in the volume and their kinetic (thermal) energy; these two factors determine the frequency of collisions between the gas molecules and the container wall per unit time, and thus the pressure. Increased temperature, which imparts a greater velocity to each gas molecule, will increase the number of collisions and thus the pressure. Similarly, if the volume is reduced by compression, there will be a greater number of molecules per unit volume adjacent to the wall, and again the probability of collision with the walls is increased. This is really only an intuitive re-statement of the ideal gas law, which quantifies the relationships.

An important extension of the gas laws is *Dalton's Law of Partial Pressures*, which states that the total pressure in a container of a mixture of gases will be the sum of the pressures of each of the component gases, or

$$P_{Total} = P_A + P_B + \cdots + P_N \qquad \text{(Eq. 2.2)}$$

This is an important physical law for physiology, since it states that each gas exerts a pressure independent of that of any other gases present, and that

TABLE 2.1
Molar Volumes of Some
Common Gases Given in
Liters at STPD[a]

Gas	Volume of 1 mole
He	22.43
Ne	22.42
A	22.39
H_2	22.43
N_2	22.40
O_2	22.39
H_2O	22.05
CO	22.40
CO_2	22.26
NH_3	22.09

[a]Reproduced with permission
from Radford (1964).

there is a direct proportionality between the composition of a mixture of gases and the pressure exerted by each. An example of the utility of this principle is provided by considering a sample of air at sea level. The total pressure exerted by the sample is ideally 760 mm Hg, and oxygen constitutes about 20.9% of the total volume. We may therefore calculate directly that the *partial pressure* of oxygen in the sample is $(760)(0.209) = 158.8$ mm Hg.

Units of pressure are unfortunately a source of considerable confusion. In the strict physical sense, pressure is defined as force applied to or distributed over a surface, and is measured as force per unit area. In the cgs system, the proper unit of pressure is the *barye*, where 1 barye $= 1$ dyne cm^{-2}, or the more familiar bar, equal to 10^6 baryes. In physiology, these units are almost never used, and more commonly pressure is expressed as mm Hg, or torr (sometimes capitalized; derived from Torricelli, a 17th-century student of Galileo and the inventor of the mercury barometer). The torr is defined as a pressure (i.e., force) sufficient to balance the force exerted by a column of mercury 1 mm high. The familiarity of measurement of pressures in this fashion, in barometers, mercury manometers, etc., seems to this author to be a powerful argument in favor of using mm Hg, and in this book all pressures will be expressed in this way. Occasionally, when very small pressures are being measured, investigators find it more convenient to measure them against a column of water or saline solution. At room temperature these can be interconverted by the using the ratio of the density of mercury to that of water:

$$(P, \text{torr}) \times 13.6 = (P, \text{mm } H_2O) \qquad \text{(Eq. 2.3)}$$

When sea water is used, the correction is 13.1, and for intermediate saline concentrations (densities), something in between. Recently the SI system has adopted the Pascal (abbreviated Pa; 1 torr = 133.322 Pa) as the unit of pressure, but it seems a clumsy unit, and does not appear to have been embraced by most physiologists. Other units of pressure measurement, and conversion factors, are given in Appendix 1.

2.3. LIQUIDS

Liquids are fluids, like gases, but the important distinction is that in a liquid the weak cohesive forces between molecules become important, such that the molecules are still relatively free to change position with respect to each other, but are restricted to an area near other molecules. A liquid thus has an intrinsic, relatively fixed volume, independent of its container. These weak cohesive forces, or *Van der Waal's forces*, vary with the particular fluid and with temperature, so that many familiar substances are gaseous at higher temperatures and liquid at lower temperatures. In water, the Van der Waal's forces are relatively strong, so that the gas/liquid transition occurs at 100°C (373.2°K), but for other fluids of physiological importance, particularly oxygen and carbon dioxide, this transition occurs far below the physiological temperature range, the Van der Waal's forces of these materials being much weaker. Another consequence of the cohesive forces of liquids is that they may be considered incompressible within the physiological pressure range.

2.4. DISSOLUTION OF GASES IN LIQUIDS

In an idealized two-phase system of a gas and a liquid in contact, if the gas is soluble in the liquid, the system will come to an equilibrium condition in which the gas pressure in the two phases is equal (Fig. 2.1). At equilibrium, by definition, the number of gas molecules leaving the gas to enter the liquid phase will just equal the number of gas molecules leaving the liquid to enter the gas. If the gas phase is a mixture of several gases, each with its own partial pressure, then at equilibrium the system will have equal total gas pressures in the liquid and gaseous phases, and the partial pressure law will also hold, such that the partial pressure of each of the components in the gaseous and liquid phases will be equal. This is another physiologically important law, since one is frequently dealing with situations that require a complete understanding of gas/liquid equilibria. To extend the previous example, if a liquid is in equilibrium with air at sea level, then the partial pressure of

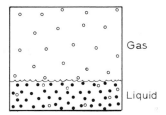

Fig. 2.1 At equilibrium the partial pressure of the gas is the same in both the gas and liquid phases. The number of gas molecules entering and leaving the liquid phase per unit time is the same.

oxygen can be calculated by knowing that if the partial pressure in the gas phase was 158.8 mm Hg, it will be the same in the liquid. This idea is depicted in general form in Fig. 2.1.

Knowing that a gas is soluble in a liquid, and what the partial pressure in the liquid will be, is not sufficient information from which to calculate the *amount*, *quantity*, or *mass* of the gaseous substance that will be present in a given volume of the liquid. For this, an additional piece of information relating the partial pressure in the liquid and the mass present is needed. The relationship is generally stated mathematically as

$$Q = \alpha P \qquad\qquad \text{(Eq. 2.4)}$$

where Q is the mass or amount, P the partial pressure, and α the *solubility coefficient*. A great deal of confusion arises at this point, however, since there are many ways in which the solubility coefficient can be expressed and lamentably little standardization in the physiological literature. Even the term *solubility coefficient* is not always used, as for example in Dejours' (1975) book, where he uses the term *capacitance* to mean the same thing. Most of the time, solubility is expressed as a quantity of gas contained in a volume of liquid under some convenient set of conditions, and the units of quantity are also chosen according to various conventions or expedience. There is unfortunately no symbol that can be used without some ambiguity; the Greek letter alpha (α) is most frequently used, but Bunsen defined this originally as the absorption coefficient, the volume of dry gas reduced to standard temperature and pressure (0°C, 760 mm Hg, dry, or STPD) present in one volume of liquid. The symbol α is encountered frequently with different definitions, which are not always clearly stated. The letter S is also used to denote the volume of dry gas present per volume of liquid under the conditions of the experiment, but this notation too is ambiguous, since S is used nowadays to mean fractional saturation (of blood, usually). Dejours uses the Greek letter beta (β) for his capacitance; various other authors have

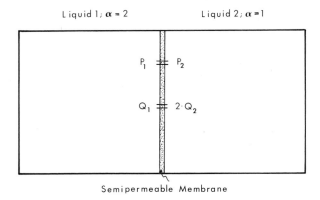

Fig. 2.2 When two liquids (1 and 2) are separated by a semipermeable membrane, the partial pressure of a gas dissolved in both will be the same; i.e., $P_1 = P_2$. If, however, the solubilities are different in the two liquids, the amounts of gas present in the two liquids will differ in direct proportion to the ratio of their solubility coefficients.

employed a potpourri of symbols, even occasionally employing α or β as non-linear functions for blood! In this book, the symbol α will be used to mean the volume of dry gas at STP contained in a liter of liquid. The use of S to denote fractional saturation will be avoided and α subscripted to indicate which gas is meant, e.g., α_{O_2} denoting the solubility of oxygen. It might, then, be desirable to re-state equation (2.4) above as

$$V = \alpha P \qquad \text{(Eq. 2.5)}$$

replacing Q for mass with V for volume, keeping in mind the relationship between volume at STP and moles of gas (i.e., $22 +$ L mol^{-1}; Table 2.1).

Just as the amounts of gas in the liquid and gas phases will be quite different at the same partial pressure, the amounts of gas present at equilibrium between two different liquids may also be different if the solubility coefficients are different. In non-equilibrium cases, we may have circumstances in which gas will diffuse *up* a concentration gradient, while diffusing *down* a partial pressure gradient. This idea is illustrated in Fig. 2.2. By setting P_1 equal to P_2,

$$V_1/\alpha_1 = V_2/\alpha_2 \qquad \text{(Eq. 2.6)}$$

In other words, although the system is at equilibrium and the partial pressures are the same, the amounts of oxygen in the two liquids are different. This point cannot be over-emphasized: equilibrium occurs at equal partial pressures, not necessarily equal concentrations. It is furthermore a common kind of problem, for example in assessing exchange between water and blood or blood and cerebrospinal fluid.

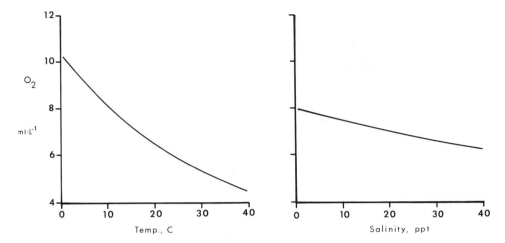

Fig. 2.3 The solubility of oxygen as a function of temperature in pure water (left) and as a function of salinity at 10°C (right). The units of solubility are ml O_2 L^{-1} equilibrated with a water-saturated atmosphere of 20.94% oxygen.

The solubility of gases is influenced by many factors, only one of which, the nature of the solvent, has been mentioned. Temperature and the solubility coefficient are inversely but not linearly related (Fig. 2.3). Increasing ionic strength reduces gas solubility, but again, not linearly.

The solubility of gases is for practical purposes insensitive to the presence of other gases and insensitive to the nature of the electrolytes present. The solubilities of some common gases in water are given in Table 2.2, and some more detailed tables of gas solubility have been provided in Appendix 2.

In the above discussion, only the physical solubility of the molecule in question has been discussed; chemical reactions that bind solute gas molecules to the solvent, and thereby increase its concentration, are not germane to the discussion of solubility, although such binding reactions are of great importance physiologically, as detailed in subsequent chapters.

2.5. DIFFUSION

The mechanics of diffusion of gases can be imagined by likening the process to what would happen in a closed room with several thousand small elastic balls in violent motion on one side of the room. In a very short time the balls would be distributed evenly, more or less, within the space of the room, as a result of their random direction and velocity, the elasticity of the collisions, and the higher probability of being deflected from an area with a high density to one of low density. An ideal gas behaves in just this fashion, so that in

TABLE 2.2
The Relative Solubility of Various Common
Gases in Water at $20°C$[a]

Gas	α	Relative solubility
He	0.0088	0.28
Ne	0.0147	0.47
A	0.0337	1.07
H_2	0.0182	0.58
N_2	0.0155	0.49
O_2	0.0314	1
CO	0.0232	0.74
CO_2	0.872	27.8
NH_3	715.4	22,783.

[a]The Bunsen coefficient (α) is defined as the
volume of gas (STPD) dissolved per volume of liquid
at a gas pressure of 1 atm and the relative solubility
as α divided by the solubility of O_2.

any situation in which the concentration (or, more properly, the activity) of a gas is unequal to start with, there will be a net movement from regions of higher concentration to those of lower concentration until dynamic equilibrium is reached. The rate at which this equilibrium is reached is a function of the initial gas concentration difference (gradient) and the *diffusion coefficient* (D) of the gas.

In the early 1800's Graham and others investigated the diffusive properties of gases, often by studying the rates at which they moved from a closed container with a pinhole in one side into an adjacent evacuated container. As a first approximation, the diffusion coefficients were estimated to be proportional to the inverse square root of the atomic mass (or density) of the gas:

$$D_A / D_B = (M_B / M_A)^{0.5} \qquad \text{(Eq. 2.7)}$$

for two gases, A and B, with mass M and diffusion coefficients D. Real gases, however, do not behave in ideal fashion, and the more modern view recognizes that the attractive forces (Van der Waal's forces) cause the diffusion rates to vary from the inverse square root rule by amounts proportional to the strength of the Van der Waal's forces. Thus, rather than finding a diffusion coefficient ratio for oxygen to carbon dioxide of $(44/32)^{0.5} = 1.17$, the ratio is actually 1.82, due to the much higher mutual attraction of CO_2 gas molecules for each other. The diffusion coefficient also depends on temperature, since the driving force is thermal (kinetic) energy, but is relatively independent of the absolute pressure, at least in the physiological range.

In considering the pressures exerted by individual gases in mixtures, we have seen that each component behaves independently, and that the pressures of each are additive (Dalton's Law). Similarly, the diffusion of individual gases in mixtures proceeds independently of other components in the mixture. In biological situations this is commonplace, since oxygen is often diffusing in one direction at the same time that carbon dioxide is diffusing in the opposite direction, and the rates of the two processes are mutually independent.

The mathematics of diffusion have been extensively studied, and it is possible, for any finite bounded space, to describe via a set of partial differential equations the net diffusional movements in three-dimensional space. Rashevsky, for example, has devoted most of a volume to the subject (1960). In biological systems the imprecision of the physical description of important parameters and the lack of facility of most biologists with such intimidating equations usually dictate a much simpler approach. A one-dimensional variant of what are loosely called *Fick's Laws* (formulated around 1855) is usually presented:

$$dQ/dt = -AD(dc/dx) \qquad \text{(Eq. 2.8)}$$

This equation can be stated: The rate of change (or net movement) of a substance Q over time is equal to the product of the area through which the substance is diffusing, A; the diffusion coefficient, D; and the concentration gradient, dc/dx. This is still not an equation with much utility, since an exact mathematical description of the concentration gradient can seldom be precisely given for biological systems. We therefore usually make the further simplifying assumption that dc/dx may be replaced by $\Delta c/x$, where Δc is simply $(c_i - c_o)$ and x a linear distance measurement. In other words, we assume that the concentration gradient is linear (Fig. 2.4). This reduces the equation to

$$dQ/dt = -AD(\Delta c/x) \qquad \text{(Eq. 2.9)}$$

The sign of the right-hand term signifies that with a greater concentration at the point of reference the net movement will be outward.

2.5.1. Diffusion of Gases in Liquids

The diffusion of gases in liquids is not very different physically from "self-diffusion," or diffusion in gaseous mixtures, but in liquids the diffusion coefficients are even more dependent upon the relative strength of Van der Waal's forces, not only among molecules of gas but between gas and solvent molecules. In aqueous solution the relative diffusion rates of O_2 and CO_2

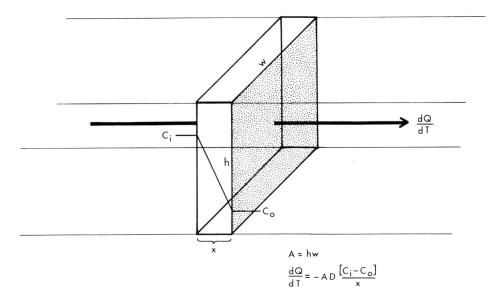

$$A = hw$$

$$\frac{dQ}{dT} = -AD\frac{[C_i - C_o]}{x}$$

Fig. 2.4 A simplified representation of diffusion in one dimension, with a linear concentration gradient through a barrier of thickness x.

are also different from the inverse square root rule and different from their ratio in gaseous media, the actual ratio being about 1.39, O_2 to CO_2. Another important difference is that the absolute rates are slower in water than in gases by about a factor of 10^5, as shown in Table 2.3.

For the example of O_2 and CO_2, although the diffusion coefficients of both are much lower in water than in gases, their ratio is not greatly different in the two media (1.17 vs. 1.39). The solubility of the two gases, however, is very different, CO_2 being about 28-fold more soluble than O_2 (Table 2.2). Since the concentration is equal to the partial pressure times the solubility, or $C = \alpha P$, the D in the diffusion equation (above) is often replaced by a new term, D', which is equal to αD, where α is the Bunsen solubility coefficient. D' is often called the *permeation coefficient*, or *Krogh's constant*, since the early work he did on diffusion dealt in these terms.

We may now re-state the diffusion equation (2.9) in a physiologically useful form, provided careful attention is paid to units. D is conventionally given in units of $cm^2\ sec^{-1}$, so that if the terms of equations (2.8) and (2.9) are stated in cm^2 for area, sec for time, and cm for distance (x), the equations are independent of units of concentration, provided the units of Q and Δc are the same. The units for α are conventionally cm^3 gas STPD $cm^{-3}\ torr^{-1}$,

TABLE 2.3
Values for the Diffusion Coefficient (D) and Krogh's Permeation Constant (D') in Various Materials[a,b]

Medium	Gas	D	D'
Air	CO_2	0.104	
	O_2	0.189	
	N_2	0.178	
	H_2	1.285	
Water	CO_2	1.77×10^{-5}	2.03×10^{-8}
	O_2	1.98×10^{-5}	8.18×10^{-10}
	NH_3	1.77×10^{-5}	1.67×10^{-5}
Muscle	O_2		3.37×10^{-10}
Connective tissue	O_2		2.77×10^{-10}
Chitin	O_2		0.31×10^{-10}

[a]The units of D are $cm^2\ sec^{-1}$, and for D' they are cm^3 gas STPD $cm^{-1}\ sec^{-1}\ torr^{-1}$ (see text).
[b]Data from the *Handbook of Physics and Chemistry*; Radford (1964); and Krogh (1919).

so the resultant D' has units of cm^3 gas STPD $sec^{-1}\ cm^{-1}\ torr^{-1}$. These conventions are unfortunately not always followed, and almost every conceivable combination of units and correspondingly different numerical values are found in the literature. With these modifications and qualifications, equation (2.8) may now be re-stated as

$$dQ/dt = -AD'(\Delta P/x) \qquad \text{(Eq. 2.10)}$$

and it is some variant or re-arrangement of this equation that is most often encountered in physiological work. It has proven to be a most useful approximation, provided the simplifying assumptions are not conveniently forgotten: one dimension only is considered; the concentration gradient has been approximated as a linear function; the permeation coefficient is different from the diffusion coefficient; and the units must be internally consistent. Precise terminology is important; it would be accurate to state that the total *permeation rate* of CO_2 in water is about 20-fold greater than that of O_2, since this is a product of similar diffusion coefficients and dissimilar solubilities.

A few additional points are worth making with regard to the diffusion and permeation coefficients in Table 2.3. The temperature sensitivity of D is about 2% per °C, but since solubility is inversely proportional to temperature and D' combines these two parameters, the result is a temperature sensitivity for D' of only about 1% per °C. A second important point is that diffusional gradients that may be effective over distances of 10 cm in air will be equally effective over distances of only 1 μm or so in aqueous media, due to the vast differences in the absolute diffusion coefficients. The aquatic animal must clearly provide much more intimate contact of the respiratory medium with the exchange surface. Diffusion in aqueous solutions is very much a short-range phenomenon.

2.6. THE ATMOSPHERE

The gas of principal interest is, of course, the atmosphere, which is a complex mixture of many gases. The most abundant are nitrogen and oxygen, but there are many minor components, including carbon dioxide, as shown in Table 2.4. At sea level, the standard atmosphere is somewhat arbitrarily assigned the pressure value of 760 torr, excluding water vapor. Water vapor is, however, an ever-present component of the atmosphere, as well as of many gases we work with. The relationship between water vapor in a saturated atmosphere and temperature is given in Appendix 3, and the proper calculation of ambient atmospheric partial pressures for various gases must take into account the relative humidity and temperature. The most common means of measurement of ambient atmospheric pressure is with a mercury barometer, in which the height of a column of mercury in an evacuated tube is measured with reference to the mercury level in a reservoir into which the tube is inserted open end down. In making barometer readings, one must make sure that the reservoir level, which can usually be adjusted, is at the proper reference mark, and one must also make the appropriate temperature and gravity corrections. The former takes into account the effect of temperature on mercury density, and the latter the variations in the earth's gravity as a function of latitude. Most barometers are supplied with correction tables for this purpose. The correct partial pressure for a given atmospheric gas may then be calculated as

$$P = P_B - (\text{R.H.})(P_W) \qquad \text{(Eq. 2.11)}$$

where P_B is the barometric pressure, R.H. the relative humidity, and P_W the water vapor pressure at ambient temperature.

2.7. CONVECTION: THE BULK FLOW OF FLUIDS

In discussing diffusive movements of matter through fluids, we have been considering the fluid matrix as fixed and unmoving. Of course this is often not the case, and more often in situations of physiological interest the fluids exhibit convection, or bulk fluid flow. At this point it is useful to give a slightly

TABLE 2.4

The Components of Atmospheric Air, Excluding
Water Vapor, at the Standard Pressure of 760
torr[a,b]

Component	% Volume	P, torr
N_2	78.084	593.4
O_2	20.946	159.2
CO_2	0.033	0.25
A	0.934	7.10
	ppm	P, millitorr
Ne	18.18	13.8
He	5.24	3.98
CH_4	2	1.5

[a]Data from the *Handbook of Chemistry and Physics*.
[b]Significant to $\pm$ 1.

more rigorous definition of a fluid as "any material unable to resist permanently even the slightest shearing force" (Burton, 1965). Solids, by contrast, resist shearing or deforming forces by virtue of their elasticity, and so long as some threshold is not reached, they will regain their original shape when the force is removed.

If flow of a fluid is considered for a planar slice oriented parallel to the direction of flow, the shearing force is tangential to the plane, and shear deformation occurs by slippage of the plane relative to parallel planes above and below it (Fig. 2.5). The fluid is not completely unlike a solid, however, since there is some resistance to flow which can be thought of as a transient elasticity, a frictional resistance between planes, or simply a "lack of slipperiness." This characteristic is *viscosity*, which varies with the fluid and the temperature.

Shear forces can arise in several ways, but in physiological systems we are generally concerned with shear forces generated mechanically, whether by the action of cilia, ventilatory apparatus, the heart, or other means. Thermal gradients may also give rise to convective flow, due to differences in density and the force of gravity acting upon the fluid, but these thermal convection processes are seldom significant in gas exchange or transport.

The physics and mathematics of fluid flow (fluid dynamics) are treated extensively in texts and specialized journals, but since some understanding of the fundamentals is necessary for the analysis of many physiological processes, a brief review follows. It is particularly useful to understand the

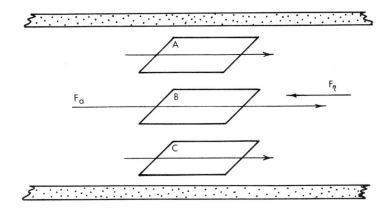

Fig. 2.5 The vector F_a represents the *shear force* applied tangential to the plane of consideration. The motion of B relative to A and C is *shear*. F_η represents the resisting force resulting from viscosity.

basics of flow velocity profiles, the relationship between pressure and flow, the calculation of work, and the conditions for laminar flow.

2.7.1. Flow Velocity Profiles

For the situation in which a fluid in contact with a surface is subjected to a shearing force parallel to that surface, whether we are considering blood in contact with vessel walls, water in contact with an external surface, or some other situation, the velocity of flow in the tangential direction will vary as a function of the distance from the surface. This is not due to any frictional resistance between the fluid and the wall material, but to cohesive forces between the molecules of the fluid and the container wall. If we consider the fluid as divided into infinitely thin planar slices, the slice immediately adjacent to the wall will have zero velocity, the next layer outward will have a very small velocity with respect to the innermost layer, the next slightly larger, and so forth. If there is a second wall (or if we have a cylinder), then the velocity will be maximal at the center and decline again as the other surface is reached, as shown in Fig. 2.6. For a cross-sectional slice of a blood vessel, the velocity profile across the lumen will have a paraboloid shape, with zero velocity at either wall. Furthermore, since the rate of change of velocity with distance in a direction perpendicular to the flow (which is shear, by definition) is least at the center, the shear is least there and greatest at the walls. The layers nearest the walls contribute the greatest *drag*, or resistance to flow.

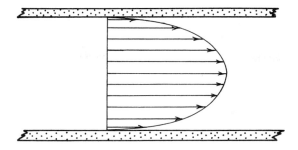

Fig. 2.6 For a shearing force applied over the area of the tube (i.e., a pressure, since $P = FA$), a paraboloid flow velocity profile results. Since dV/dx increases away from the center, the viscous drag is contributed more by layers (lamina) near the walls.

It is not, then, any particular property of the wall itself that imparts the viscous drag, but an intrinsic property of fluid flow.

Since the proportion of the fluid that is near the wall is relatively much greater in a small vessel or channel than in a large one, it is also reasonable to expect that resistance to flow will be greater in the small vessels. Pioneer work by Poiseuille in the early 1800's and later mathematical contributions by Hagen have given us the familiar equation describing this relationship for tubular channels:

$$V = (\Delta P \pi R^4)/(8\eta L) \qquad \text{(Eq. 2.12)}$$

where V is volume per second in cm^3; ΔP the difference in pressure over the length of the tube in dynes cm^{-2}; R the radius of the tube in cm; L the length of the tube in cm; and η the viscosity of the fluid in poises ($\equiv$ dynes sec cm^{-2}). Several interesting conclusions follow from this equation:

1. Pressure and flow are linearly related, given constant radius and viscosity.
2. Very small increases (or decreases) in radius cause very large increases (or decreases) in flow, provided pressure and viscosity remain the same.
3. Constant flow can be maintained in the face of falling pressure by very small increases in radius.

Since the radius is raised to the fourth power, the flow through a 1 cm radius tube would increase by a factor of 2 with an increase in radius to only 1.19 cm, and would decrease by half with a reduction to 0.84 cm at constant pressure and viscosity.

The viscosity of most fluids at constant temperature may be considered

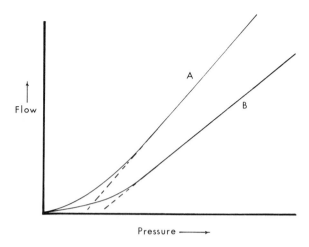

Fig. 2.7 Idealized pressure–flow relationships for blood (B) and protein (A), liquid systems which display anomalous flow behavior. The non-linearity implies changing viscosity as pressure and flow increase.

a constant, which is to say that the fluids are Newtonian in behavior. Certain fluids of biological importance, however, exhibit significantly non-linear viscosity behavior, usually caused by dissolved (colloidal) proteins. The relationships between pressure and flow for a hypothetical protein solution and for blood are shown in Fig. 2.7.

2.7.2. Laminar and Turbulent Flow

The flow velocity profile shown in Fig. 2.6 applies to conditions under which flow may be considered in layers, planar slices, or concentric cylinders and is termed *laminar flow*. Under these conditions, the applied force is translated entirely into shear in the direction tangential to the plane of the slices or cylinders. At higher flow velocities, however, a new element enters in: the formation of eddies and other flow currents in directions other than the tangential (Fig. 2.8). The velocity at which the transition from laminar to turbulent flow occurs varies with the size of the flow channel and with the viscosity of the fluid, higher values of each favoring turbulence. Bends in the flow channels, and sometimes surface roughness, may also lead to turbulence. Because of the excess energy that must be translated into the turbulent component, the transition from laminar to turbulent flow is seen as a break in the linear pressure/flow relationship, such that increases in pressure produce proportionately less increase in flow (Fig. 2.9).

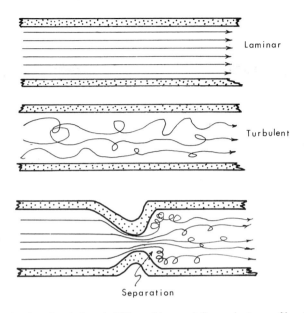

Fig. 2.8 Laminar (top) and turbulent (middle and bottom) flow velocity profiles. Turbulent flow is usually dominant at high flow rates but may be induced at lower flows by irregularities in the wall (lower).

Turbulence is infrequent in biological situations, especially in the aquatic animals. Examples from mammalian physiology are probably familiar; perhaps the most dramatic is the production of so-called Korotkow sounds by turbulence around the heart valves and in the aorta of man. There do not appear to have been any investigations of the presence of heart sounds in the aquatic animals, but except for some highly active animals such as tunas, it is doubtful that high enough velocities are ever achieved in the circulatory or respiratory systems. Turbulent flow dynamics become quite important in the consideration of swimming and flying drag, but these subjects have been reviewed extensively (Webb, 1978) and are outside the scope of this book.

2.7.3. The Unstirred Layer

In the literature on both gas exchange and ion exchange, the term *unstirred layer* is frequently encountered. It is clear from the foregoing discussion that this term is a little misleading, since it seems to imply that there is some layer of finite thickness in which flow is zero, bordered by some layer farther out that is completely stirred or has a high velocity.

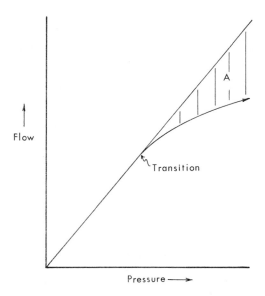

Fig. 2.9   Above the laminar-to-turbulent flow transition, less flow results for equal increments in pressure. In region A, an increasing amount of the driving energy is going into turbulence eddies, rather than linearly directed flow.

In fact, the flow velocity proceeding outward from a boundary wall increases in a progressive and continuous fashion. What is generally meant by the term *unstirred layer* is that area near a surface in which convective flow is insignificantly low and in which pure diffusion is the rate-limiting process. The concept of an unstirred layer is nonetheless important, however, since steep gradients may arise in this low-convection zone, and the addition of this thickness to the diffusion path length (cf. Eq. 2.10) has a profound influence on the analysis of membrane transfer problems.

2.7.4. The Work of Pressure and Flow

The definition of work from physics is the energy expended by a force acting over a distance, $W = Fx$. For the example of flow and pressure, we have seen that pressure is a force applied over an area, or $P = F/A$, and if we substitute a pressure term in the work equation, we have

$$W = PAx \qquad \text{(Eq. 2.13)}$$

By combining A and x we get units of volume (cm^3), and dividing both sides by a term for time, we arrive at the useful equation

$$W/t = PV/t \qquad \text{(Eq. 2.14)}$$

where the units of W are ergs (dyne cm), t is in seconds, P in dynes cm^{-2}, and V in cm^3. This equation is adequate for describing the work done in steady flow at constant pressure, but more often in biological situations the pressure and flow are continuously variable in time, so a slightly more complex equation is needed

$$\dot{W} = \int PV \, dt \qquad \text{(Eq. 2.15)}$$

where $\dot{W}$ is the rate of work per unit time and the pressure–volume relationship is integrated over the time period. One seldom has an explicit mathematical expression for the pressure/volume pattern, so the equation is usually solved by various expedients. One useful method is to estimate the area graphically, either with an integrating planimeter or by the "cut and weigh" method.

One important aspect of the work calculation that may be easily overlooked is that the calculation provides a numerical estimate of the physical work performed on the fluid, sometimes called the *external work*. This is not the same as the total work done by the organ propelling the fluid; the calculation of this total metabolic work by the organ must include the efficiency of the organ in translating metabolic work into mechanical work. Even the heart, which is a structurally efficient organ compared to many others, has an efficiency below 10% under normal light-load conditions, so the total work done by an organ to bring about convective flow can easily be an order of magnitude greater than the external work that results.

2.8. THE PROPERTIES OF WATER AND AQUEOUS SOLUTIONS

Life itself depends on the highly anomalous characteristics of water as a solvent. The water molecule is highly polar, which gives it a high dielectric constant, a considerable tendency to bind not only other water molecules but also solute molecules, a high specific heat, and a density minimum at 4°C. The last property dictates that ice will form on the surface of bodies of water in winter, thereby protecting deeper-lying waters from freezing completely. The various properties of water also give it a very high surface tension, a high wetting tendency, and high capillarity; each of these properties is important in maintaining cellular integrity. Some of the salient physical and chemical characteristics of water are given in Table 2.5.

Several of these properties of water are influenced by solvent particles dissolved in it, and these solute-dependent properties are termed *colligative* properties. The addition of solutes like NaCl to water were observed long ago to lower the freezing point, raise the boiling point, and decrease the vapor

TABLE 2.5
Characteristics of Water[a]

Property	Value
Density at 0°C	0.999968 g ml^{-1}
At 3.98°C	1.000000 g ml^{-1}
At 20°C	0.998234 g ml^{-1}
Thermal conductivity	5.14 cal hr^{-1} cm^{-1} °K^{-1} at 20°C
Viscosity at 0°C	1.79 centipoise
At 20°C	1.002
At 40°C	0.65
At 100°C	0.28
Specific heat	0.9988 cal g^{-1} °C^{-1}
Dielectric constant	80.1
Surface tension	72.8 dynes cm^{-1}

[a]Values given are for 20°C unless otherwise stated.

pressure of water. In the 1880's Raoult observed that the change in these properties was a direct function of the molar quantity of solute added. Through various refinements worked out by Arrhenius, Ostwald, Van't Hoff, Debye, and Hückel, we now have a comprehensive body of theory to explain the quantitative relationship between the changes in colligative properties of water and the amount and nature of solutes added to it. The simplified theory predicts that the changes will be proportional to the number of particles in the solution, so that a completely dissociable solute like NaCl would add 2 moles (i.e., two times Avogadro's number of particles) to the solution when completely dissociated. The addition of 1 mole of active particles to water increases the boiling point by 0.54°C, and depresses the freezing point by 1.86°C, and a solution that changes these values by those amounts is said to be 1 *Osmolar*.

2.8.1. Activity vs. Concentration

Early investigations of such solution properties led to some apparent anomalies. For example, the addition of 1 mole of NaCl to pure water should produce 2 Osm of ions when completely dissociated into Na^+ and Cl^-, and should therefore lower the freezing point by 1.86°C. The actual reduction is 1.58°C, which was originally interpreted to mean that the NaCl was only 85% dissociated.

We now know that very dilute solutions of non-electrolytes, i.e., less than 0.2 *M*, behave in a reasonably ideal way, but dilute solutions of electrolytes, which include most physiological fluids, deviate significantly from ideal behavior. Ions in solutions are subject to the influence of attractive forces between solute and solvent and to the effects of electrical fields on movement of charged particles. In the early 1900's, Gilbert Lewis described a thermodynamic theory which accounted for these and other solution anomalies. He derived two parameters, the *fugacity* and *activity* of a species in solution. The fugacity, literally an escaping tendency, is defined as a change in free energy compared to a reference state, and is a general property of any chemical species in any phase system. The activity is defined as the ratio of the fugacity in a particular state to the fugacity of the reference state. Activity expresses a true "chemical potential" which takes into account the non-ideal influences of interionic attraction, etc.

The strength of the various intrasolution forces is dependent on the concentration of each solute and on the total solutes in solution, and so the relationship between activity and concentration is complex. For most chemical equilibria of the type

$$AC \longleftrightarrow A^- + C^+ \qquad \text{(Eq. 2.16)}$$

the true equilibrium constant K should be calculated using activities of each reaction species, rather than concentrations. In most physiological systems, however, we work instead with the apparent dissociation constant K', which incorporates the activity coefficients; K' can be derived from the value for K and the *ionic strength* μ by the relationship worked out by Debye and Hückel:

$$K' = K \pm 0.52(\mu)^{1/2} \qquad \text{(Eq. 2.17)}$$

Ionic strength is defined as

$$\mu = \tfrac{1}{2}(C_1 z_1^2 + C_2 z_2^2 + \cdots) \qquad \text{(Eq. 2.18)}$$

where C_i is the concentration of the ith species, and z_i is its valence.

The case of the H^+ ion is a particularly important one, discussed in more detail further on. In this case, however, it should be pointed out that the activity coefficient for H^+ conventionally takes into account not only the non-ideal solution properties discussed here but also some arbitrary effects of the potentiometric standards employed in the pH scale. Part of the reason for this is the impossibility of determining the absolute activity coefficient for any single ionic species in a complex solution.

The activity coefficients for a variety of common solutes are found in Table 2.6.

TABLE 2.6
Activity Coefficients for
Some Common Ionic Species
in Aqueous Solutions[a,b]

Compound	Activity coefficient
HCl	0.796
HNO_3	0.791
KCl	0.770
KNO_3	0.739
KOH	0.798
LiCl	0.790
$MgSO_4$	0.150
NaCl	0.778
NaH_2SO_4	0.744
NaOH	0.766

[a]Data from the *Handbook of Chemistry and Physics*; Hills (1973).
[b]All values are for 25°C and a concentration of 0.1 molal.

2.8.2. Osmosis

When two solutions of differing ionic strength are separated by a semipermeable membrane, i.e., one permeable to solvent water but not to solute ions, water will tend to move from the solution where its activity is lower to the solution where it is higher. In other words, there will be a net movement of water toward the solution with higher ionic strength, and the movement will continue until the concentrations are equal. The net movement may be opposed by a hydrostatic pressure, and the pressure necessary to stop net water movement from pure water to a given solution is termed the *osmotic pressure* of the solution. In general, the relationship of osmotic pressure to concentration is given by

$$aC = (O.P.)/RT \qquad \text{(Eq. 2.19)}$$

where C is the concentration in moles L^{-1}, R and T have their usual meanings, and O.P. is the osmotic pressure in Osmoles. Thus the addition of 1 mole of active particles (1 Osmole by definition) will generate an osmotic pressure of 22.4 atm.

2.8.3. Sea Water

Sea water is a solution of particular biological interest, not only because so many of the extant animals inhabit it, but because it was almost certainly the milieu which gave rise to the first life forms. The ionic composition of

TABLE 2.7
Some Chemical and Physical Properties of Sea Water[a]

Property	Value at 20°C
Density	1.02476 g cm^{-3}
Conductivity	47.92 (ohm-cm)$^{-1}$
Refractive index	6463
Sound speed	1521 m sec^{-1}
Molecular viscosity	1.09 cp
Specific heat	3.9937 J g^{-1} °C^{-1}

Major elements in solution g kg^{-1}

Cl	18.98
Na	10.56
Mg	1.27
S	0.884
Ca	0.40
K	0.38
Br	0.065
C (inorganic)	0.028
Sr	0.013
B	0.005 (mg kg^{-1})
Si	0.02–4.0
N (inorganic)	0.001–0.7
N (organic)	0.03–0.2
P (inorganic)	0.001–0.10
Ba	0.05
I	0.05

[a]Data from the *Handbook of Chemistry and Physics* and miscellaneous sources.

sea water is known to vary little over the world's oceans; though the total salt content (or salinity) may vary, the ratios of the various dissolved ions are for all practical purposes constant. Some of the physical and chemical characteristics of sea water are given in Table 2.7.

LITERATURE CITED

Burton, A. C. 1965. Physiology and Biophysics of the Circulation. Year Book Medical Publishers, Chicago.

Dejours, P. 1975. Principles of Comparative Respiratory Physiology. Elsevier North-Holland, New York. 253 pp.

Hills, A. G. 1973. Acid–Base Balance: Chemistry, Physiology, Pathophysiology. Williams & Wilkins, Baltimore, 381 pp.

Rashevsky, N. 1960. Mathematical Biophysics: Physico-Mathematical Foundations of Biology. Dover Publications, New York. 488 pp.

Weast, A. C. 1971. Handbook of Chemistry and Physics. CRC Press, Cleveland, Ohio.

Webb, P. W. 1978. Hydrodynamics: nonscombroid fish. *In* Fish Physiology, eds. W. S. Hoar and D. J. Randall. Academic Press, New York. Vol. VII, pp. 190–237.

SUGGESTED FURTHER READING

Radford, E. P. 1964. The physics of gases. *In* Handbook of Physiology, Sect. 3, Vol. I. Respiration. (W. O. Fenn and H. Rahn, eds). Amer. Physiol. Soc., Washington, D.C. pp. 125–152.

CHAPTER

3

BASIC ELECTRONICS FOR MEASUREMENT

3.1. INTRODUCTION

In some kinds of work the physiologist is concerned directly with measuring electrical parameters, such as the resting voltage across a membrane or the patterns of nerve discharges. More often, however, the event or parameter of interest is physical, mechanical, or chemical in nature, and the measurement involves *transducing* the information to an electrical signal that can be measured and recorded. A transducer is defined as any device that takes an input signal and converts it to an output signal of a different form. A simple example would be the mercury thermometer, which accepts an input (heat) and transduces it to a visible physical form (the length of the mercury column). For many kinds of information gathering the desired output of the transducer is an electrical signal, and most of the measurement instruments in use are based upon electrical transduction of one sort or another.

In order to take full advantage of the power and versatility of the electronic instrumentation available today, it is not necessary to have formal training in electrical engineering or any sort of electronics. On the contrary, the emphasis in commercial laboratory instrument design is on ensuring that such knowledge is not required for routine operation of the instrument. Neither is it necessary to know how to repair malfunctions of the equipment; we rely on skilled technicians for these services. On the other hand, it is usually a great advantage to have an understanding of the *principles* of operation

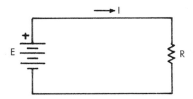

Fig. 3.1 A simple circuit with a voltage (E) supplied by a battery, containing only one resistor (R), producing a flow of current (I).

of the instrument, including a general idea of its functional organization. There is an intangible side to this, as well as a practical one: scientists tend to shy away from the use of instruments they do not understand, and may thereby limit their analytical powers due to lack of familiarity.

In the following sections, the principles of the basic electrical circuit are briefly reviewed. The emphasis throughout is on the kinds of things that are useful to know in using scientific instruments of various sorts. Some simple circuits are given that are useful in modifying signals in order to measure them more conveniently. Some common electrical test and measurement instruments are discussed, although the details of operation (the "knob twiddling") are left for the manufacturer's instruction manuals. Finally, there is a brief discussion of digital techniques and microcomputer interfacing.

3.2. SIMPLE CIRCUITS

The simplest sort of electrical circuit is shown in Fig. 3.1, a circuit in which current from a battery flows through a resistor and returns to the battery. (Common electrical symbols are given in Appendix 3.1). The voltage, current, and resistance of such a circuit are related by Ohm's Law:

$$E = IR \qquad \text{(Eq. 3.1)}$$

where E is the voltage in volts (also indicated by V), I the current in amperes (amp or A), and R the resistance in ohms (symbol Ω). For the example in Fig. 3.1, a battery voltage of 3 V and a resistor value of 10 Ω will produce a current of 0.3 A. The power dissipated in this circuit may be calculated from

$$P = IE \qquad \text{(Eq. 3.2)}$$

where P is the power in watts (symbol W). By substituting the right-hand term from (3.1), we also see that

$$P = I^2R \qquad \text{(Eq. 3.3)}$$

and the power dissipation for the circuit in Fig. 3.1 is 0.9 W.

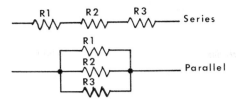

Fig. 3.2 (a) Series resistance. (b) Parallel resistance. See the text for relevant equations.

3.2.1. Resistance

An electrical element that allows a current of 1 A to flow with an applied voltage of 1 V has by definition a resistance of 1 ohm (Ω)(Eq. 3.1). The resistance of a given element is a function not only of its chemical composition but also of its cross-sectional area; thus, small wires have higher resistance than large ones. This explains the need for heavy–gauge wires connecting heaters and other equipment that carry high current: if long, small wires were used, they would dissipate appreciable power and would heat up to an unacceptable degree.

Resistor materials employed in measurement circuits are of many different materials, but one characteristic of particular importance in measurement applications is the temperature coefficient. The most common type of resistor (and the cheapest) is made with carbon, a material with a high negative temperature coefficient. That is, as the temperature rises, the resistance of a carbon resistor decreases. Several types of resistors with low temperature coefficients are available, such as the metal film and "Cermet" types, and for measurement circuits where stability is required, these should be used in preference to carbon.

Practical circuits seldom have a single resistor, and when several resistors are present in a circuit, they may be connected either in *series* or in *parallel* (Fig. 3.2). For series resistance, the total resistance is simply the sum of the individual resistances:

$$R_T = R_1 + R_2 + R_3 \qquad \text{(Eq. 3.4)}$$

using the example in Fig. 3.2. The *voltage drop* across each resistor may also be calculated from Ohm's Law, since it applies to parts of circuits as well as to the whole circuit. The total circuit current must pass through each resistance equally; so, for resistance R_1:

$$\Delta E_1 = IR_1 \qquad \text{(Eq. 3.5)}$$

and similarly for each resistance in the circuit.

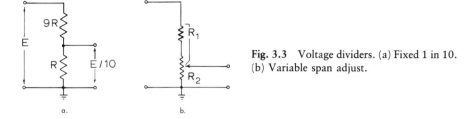

Fig. 3.3 Voltage dividers. (a) Fixed 1 in 10.
(b) Variable span adjust.

For circuit resistances arranged in parallel, as in Fig. 3.2, the total resistance is calculated as the reciprocal of the sum of the reciprocals, or

$$1/R_T = 1/R_1 + 1/R_2 + 1/R_3 \qquad \text{(Eq. 3.6)}$$

In this case the voltage drop across each resistor is the same, but the current will vary and may be calculated for each resistance by the application of Ohm's Law.

One of the simplest and most useful devices for processing electrical information is the *voltage divider*, which is based upon the voltage drop relationship for series resistances. Let us say, for example, that one wishes to record a signal that varies between 0 and 1 V, but the only recording or measuring device available has a full scale range of 0 to 100 mV (millivolts). The voltage divider circuit shown at the left in Fig. 3.3a, constructed from two fixed resistors, will provide a 1 to 10 reduction of the input signal. For accommodation of a variety of input ranges, the circuit shown at the right in Fig. 3.3b may be used, consisting of a single variable resistor, or often of a variable and a fixed resistor in series to provide finer adjustment. By simply adjusting the ratio of R_1 to $(R_1 + R_2)$, any desired input signal reduction may be achieved. The total resistance value, however, must be chosen with care and with some knowledge of the current capability of the input signal. If resistances of 9 and 1 ohms were used, for example, the current flowing in that circuit at 1 V input would be 100 mA, which may be too great for the input source to generate. On the other hand, if values of 9 and 1 megohms (10^6 ohms) were used, the recording device connected across R_1 may have an input resistance low enough to significantly reduce the total parallel resistance of R^1 and the recording device, thus changing the divider ratio. The input resistance of recording and measuring devices is discussed further in Section 3.3.

Another simple application of resistors is shown in Fig. 3.4. In some applications, the input signal is in the form of a varying current, but it may be desirable to convert it to a voltage for measurement. In that case, application of Ohm's Law shows that the voltage drop across a resistor will

Fig. 3.4 Current to voltage conversion with a resistor. The current (I) flowing through the resistor (R) produces a voltage drop that can be measured at the points indicated.

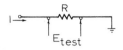

vary in proportion to the current flowing through it. In Fig. 3.4, the current signal path is completed with a resistor and the measuring device attached across the resistor. The resistor chosen should usually have low resistance value (say, 1 to 10 ohms) and temperature stability, such as a precision metal film type. The value chosen will determine the voltage range, but it should not be high enough to prevent the maximum current flow at the voltage of the input signal.

3.2.2. Capacitance

In order to understand capacitance, it is perhaps useful to visualize the physical structure of a capacitor, shown in simplified form in Fig. 3.5. The simplest form of a capacitor consists of two metal plates separated by a small air gap. At the moment of switch closure in the circuit shown, some electrons tend to be pulled off the upper plate toward the positive battery pole, and excess electrons pile up on the negative plate, generating an electric field across the two plates at the battery potential. If the two plates are discharged by the momentary connection of a wire between them, the current flow will be proportional to the *capacitance*. In practice, capacitors are usually constructed with many layers of alternating plates, the plates usually consisting of an etched metal foil or other thin conducting layer and the air replaced by any of a large variety of materials. The total capacitance is a function of the plate surface area, the separation between the plates, and the insulating material. The capacitance developed with air as the insulating material is arbitrarily assigned a value of 1, and the ratio of the capacitance developed by other materials to that with air is called the *dielectric constant* of the material. Polystyrene has a dielectric constant of 2.6, mica 5.8, and so on; many different materials are employed in capacitor construction, and the choice for any given application depends on the total capacitance needed, longevity, cost, temperature stability, high voltage resistance, etc. Most of this does not concern us here, but it should be noted that some types are polarized and must be inserted into circuits with proper polarity, while others may be connected in either direction. Connection of a large electrolytic capacitor with reversed polarity can result in an explosion.

The fundamental unit of capacitance is the *farad*, which is defined as the

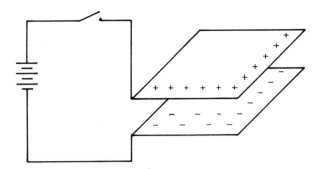

Fig. 3.5 Diagram of the simplest capacitor circuit, consisting of two metal plates separated by an air gap. When the switch is closed, charge accumulates on the plates in the manner shown.

capacitance produced by a potential of 1 V when charged by a current of 1 amp-sec. In practice, this unit is too large to be of much use (a 1 farad capacitor is as big as a bucket), and one commonly encounters units of microfarads (μF) and picofarads (pF) in practical circuits.

Capacitors arranged in parallel follow the reverse of the rules that apply to resistors. That is, the total capacitance for parallel arrangements (as in Fig. 3.6a) may be calculated as the simple sum of the individual values:

$$C_T = C_1 + C_2 + C_3 \qquad \text{(Eq. 3.7)}$$

and for a series arrangement (Fig. 3.6b) the reciprocal sum of reciprocals rule applies:

$$1/C_T = 1/C_1 + 1/C_2 + 1/C_3 \qquad \text{(Eq. 3.8)}$$

Note that for series arrangements (as with parallel resistors) the total capacitance will always be smaller than the smallest value capacitor in the circuit.

3.2.3. Time Constant

In the discussion of the function of a capacitor, only the transient current flow after the switch is turned on (Fig. 3.5) was mentioned. No direct current will flow through an ideal charged capacitor (though there is some small current in real ones); it is the rate of charging and discharging that is the useful characteristic of capacitors. For a simple circuit containing resistance

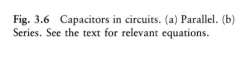

Fig. 3.6 Capacitors in circuits. (a) Parallel. (b) Series. See the text for relevant equations.

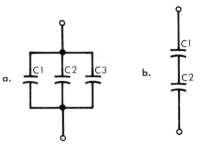

Fig. 3.7 An RC circuit with a switch and battery.

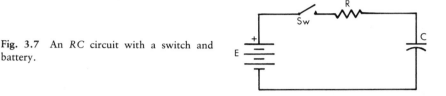

and capacitance, as in Fig. 3.7, when the switch is first closed, the charge on the capacitor is zero and the rate of current flow is determined by the value of the resistor. As the capacitor's charge increases, the voltage drop across the resistor becomes progressively less, and so the current flow drops until, at full capacitor charge, current flow is zero. The relationship between charge and time is given by the following simple exponential:

$$E(t) = e^{-kt} \qquad \text{(Eq. 3.9)}$$

At $t = 0$, the value of this expression is 1, and at a time, t, equal to $1/k$, the value is $1/e$, or 0.368. The change in value, then, in $1/k$ seconds, is 63.2%, and that amount of time is defined as the *time constant* (Υ). The time constant for the circuit shown in Fig. 3.7 is found from the simple relationship

$$\Upsilon = RC \qquad \text{(Eq. 3.10)}$$

where R is given in ohms and C in farads. For the circuit values shown, $\Upsilon = (100{,}000)(0.000001) = 0.1$ sec, and the capacitor charge as a function of time will be that shown in Fig. 3.8. The relationship between the percentage completion of an event and the number of time constants elapsed is given in Table 3.1, from which it can be seen that an event may be considered virtually complete after four or five time constants, but the relationship is asymptotic and theoretically never reaches 100%.

The relationship between current flow and time after switch closure in the

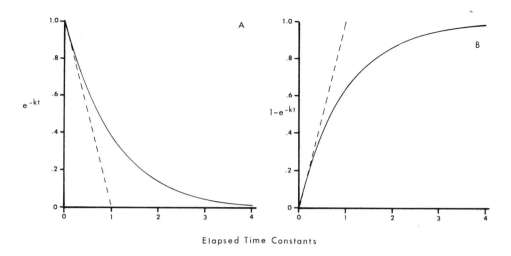

Elapsed Time Constants

Fig. 3.8 Time constant decay (a) and rise (b). See the text for equations and discussion.

circuit of Fig. 3.7 is given by the following increasing exponential function:

$$E(t) = 1 - e^{-kt} \qquad \text{(Eq. 3.11)}$$

The time constant for this relationship is calculated the same way, and the time course of current flow is shown in Fig. 3.8b.

Instead of a simple switch closure circuit, we may visualize the responses to step changes of a DC potential. The response of the circuit would be the same as the response from zero potential, and the time course of charge and current change would follow the same pattern. This leads to another very simple but useful circuit, shown in Fig. 3.9, for filtering, or damping noise in signals of interest. Let us suppose we had an electrical signal with considerable noise, but were only interested in slow changes, on the order

TABLE 3.1
The Percentage Completion
of a Transient vs. the Number
of Elapsed Time Constants

n	% Complete
1	63.2
2	86.5
3	95.0
4	98.2
5	99.3
10	99.996

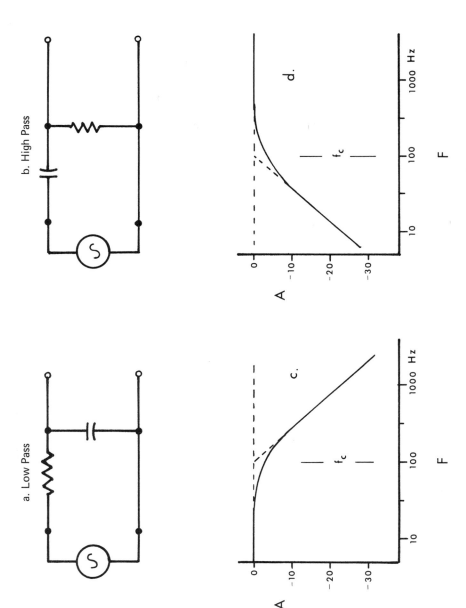

a. Low Pass

b. High Pass

c.

d.

Fig. 3.9   RC Filters. (a) Low pass. (b) High pass. (c) Low pass attenuation. (d) High pass attenuation.

of 1 sec or so. Connection of a 100 kohm resistor and a 10 μF capacitor, as shown at the left in Fig. 3.9a, to the input terminal of our measuring device would damp out signals of high frequency, since it would respond to changes with a time constant of 1 sec. The damping of high frequencies follows the curve shown in Fig. 3.9a and does not provide a sharp cut-off, but rather a logarithmically increasing attenuation as frequency rises. For sharper cut-off, one of a variety of "active" filter designs must be used, the design of which can become quite complex.

The simple RC filter can also be configured as a high pass filter, as shown at the right in Fig. 3.9b. The time constant is again simply RC and the cut-off frequency is given by

$$f_x = 1/(2\pi RC) \qquad \text{(Eq. 3.12)}$$

as for the low pass filter. The frequency attenuation is shown in Fig. 3.9b and has the same gradual "roll-off."

3.2.4. Inductance; Transformers

When current flows through any conducting material, an electrical field surrounds the conductor. At the instant when current begins to flow, some amount of the electrical energy must be subtracted from the current flow for transfer to the electrical field. This energy requirement for field generation appears as an opposing voltage, and is proportional to the rate of change of the current and to a property of the material in which the field is generated, called the *inductance* of the material.

Besides generating an electric field from the changing voltage in the conductor, the changing field will induce a varying voltage in a second conductive material within the field. This is called *mutual inductance* and concerns the physiologist in two ways: in the generation of unwanted noise and in the operation of transformers. One common form of noise pickup is induced 60 Hz noise from normal line voltage. This type of noise is usually taken into account in the design of electrical instrumentation, but the best designs of the engineers are thwarted if long signal input wires are laid near AC power cords. Since field strength falls off rapidly with increasing distance, it is usually sufficient to move these input and power leads far apart.

In transformers, the most common arrangement is to have two coils of wire, the *primary* and *secondary* coils, wrapped around a common iron core (Fig. 3.10). The field induced by the primary winding produces a voltage

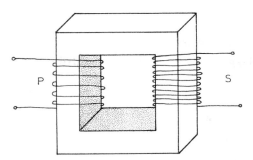

Fig. 3.10 Diagram of a transformer. It consists of a primary winding (P) and a secondary winding (S) around a common core, usually iron.

in the secondary by mutual inductance. The voltages of the two windings are determined by the ratio of the number of turns on the two coils:

$$E_s = E_p(n_s/n_p) \qquad \text{(Eq. 3.13)}$$

where s and p denote secondary and primary and n is the number of turns for each. Thus, by winding 1000 turns on the secondary and 10 on the primary, an AC voltage of 10 V could be boosted to 1000 V on the secondary. The current capability, however, is reduced by the reverse ratio, so that 1 A of primary current would produce only 0.01 A of secondary current at the higher voltage. A common application of transformers is in power supplies, where the transformer is employed to derive the voltages required to run various electrical apparatus from the 110 (or 220) V AC line voltage.

3.2.5. Impedance

The term *impedance* (symbol Z) is similar to the simpler term *resistance*, except that in some circuits the hindrance to AC current flow may also have *reactive* as well as resistive components. A full discussion of reactance may be found in any basic electronics text, but in most physiological measurements the frequencies of signals are low and circuit impedance is mainly pure resistance. The more general term *impedance* is often used, however, in stating the input characteristics of many electrical measurement devices, and as a general rule of thumb, the input impedance of the measuring device should be at least two orders of magnitude (preferably three) greater than the output impedance (or current sourcing ability) of the signal to be measured.

3.3. AMPLIFICATION

Any device capable of increasing the magnitude or power level of an electrical signal is an *amplifier*. The needs for amplifiers of one kind or another in physiology are many, so a brief discussion seems warranted. Amplifiers may be used for two main purposes: increasing the voltage or current level of small electrical signals and processing the wave forms of an input signal to a more desirable form. Amplifiers used for magnification are built into most electrical instrumentation, and the functioning of these integral units is of no particular concern. Often, however, an additional amplifier stage is required between the signal of interest and the measurement or recording device available. A simple, albeit expensive, solution is to process the signal through a commercially available *pre*amplifier unit made for the purpose at hand. Often too, some additional function is required from the amplifier, such as filtration of noise or isolation from ground, such as in electrical recordings from nerves, in which an isolation amplifier is usually employed.

In many applications, however, it is convenient to have a working knowledge of the application of the *operational amplifier*. This term, now shortened to "op amp," was borrowed originally from the analog computer field and refers to an amplifier with high input impedance, low output impedance, and high voltage gain. Op amps are available in many different integrated circuit packages at costs as low as 45 cents, and have become general-purpose amplifying devices. The typical op amp requires positive and negative regulated supply voltages, usually ± 12 or 15 V, has an inverting and a non-inverting input, and an output that may be coupled to the input by a feedback resistor to provide accurate control of gain. Some very low-cost op amps with extra high input impedance are also available that will serve quite well as electrometer-type amplifiers; the AD515 (Analog Devices, Inc.), for example, has an input impedance of around 10^{13} ohms.

3.3.1. Inverting Amplifier

The basic configuration of the inverting amplifier is shown in Fig. 3.11, with pin numbers given for the industry standard 741 type. Many other "chips," which may be selected for greater input impedance, lower zero offset, higher stability, etc., use the same pin pattern, but even if the pin arrangement is different, the principles are generally the same. The *gain* (the ratio of output to input voltage or current) of the inverting amplifier is set by the ratio of the feedback resistor, R_b, to the input resistor, R_a, and has a practical limit of 1 to 10,000 for different op amps. Due to design details of the op amp's construction, there will often be an input bias current, i.e., a current drawn at the input, and amplified and added to the output. This bias current may

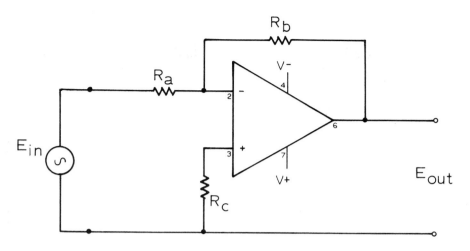

Fig. 3.11 An op amp circuit configured as an inverting amplifier. The output (E_{out}) equals $-E_{in}(R_b/R_a)$. The resistor R_c compensates for input bias current, and should be about equal to the parallel value of R_a and R_b (see Eq. 3.6). The pin numbers shown are for the industry standard 741 type; many others have the same pin arrangement. The positive and negative supply voltages ($-V_s$ and $+V_s$) are usually ±15 VDC.

be balanced by adding the optional resistor R_c, choosing a value equal to the parallel value of R_b and R_a (see Section 3.2.1). There may also be a small zero offset, i.e., a voltage that appears at the output with zero input. Most op amps have a provision for adding a balancing potentiometer to null this offset voltage.

3.3.2. Non-Inverting Amplifiers

The circuit arrangement for a non-inverting amplifier using the 741 op amp, as shown in Fig. 3.12a, is not much different from the inverting case. The feedback resistor is now connected to the inverting input, and the inverting input is tied to ground through an input resistor, R_a. The gain is determined as

$$E_{out} = E_{in}(R_b + R_a)/R_a \qquad \text{(Eq. 3.14)}$$

A special case of the non-inverting op amp is the *voltage follower*, shown in Fig. 3.12b. A typical use of this circuit would be as a preamplifier for a pH electrode or microelectrode whose output impedance is very high (i.e., has a very small current-generating capability). The voltage follower usually has a very small input current requirement (high input impedance) and a gain of 1 with a very low output impedance, suitable for further measurement

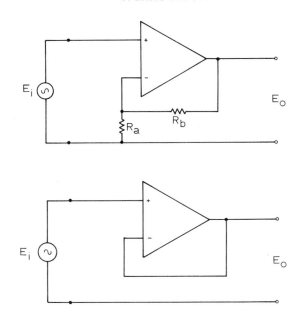

Fig. 3.12 (a) A non-inverting op amp circuit. The gain is equal to $(R_b + R_a)/R_a$. (b) A non-inverting voltage follower circuit; the gain is 1.

or signal processing. The AD515 (Analog Devices) or CA3140 (RCA) are good circuits to use, and a typical application might be to mount them on a probe directly adjacent to a remote electrode. This avoids stray noise pickup that inevitably occurs when very small signals are transmitted through long leads.

3.3.3. Differential Amplifiers

The two circuit configurations above assume that one side of the input signal is grounded, i.e., that the input signals are "single-ended." If this is not the case and the difference between two voltage levels is to be measured, the differential amplifier circuit shown in Fig. 3.13 may be used. If the ratios of resistors $R_b : R_a$ and $R_b' : R_a'$ are kept equal, the gain is simply R_b/R_a. In practice, this configuration is a little trickier, since imbalance in the resistors or drift with age or temperature will cause the gain and offset to drift.

A useful type of differential amplifier is the instrumentation amplifier. These integrated circuit packages have high impedance differential inputs, provisions for variable or programmable gain, zero offset, etc., and may be purchased for $10 to $50, depending on performance grade and features. The AD521

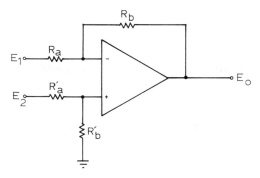

Fig. 3.13   An op amp in a differential configuration. $R_a = R'_a$, and $R_b = R'_b$, and the gain is R_b/R_a. Small imbalances in the resistor values may create large errors; matched precision resistors or small trimmers should be used.

and AD524 from Analog Devices, for example, provide gain from 1 to 1000 and have excellent *common mode* rejection. This means that varying signals present on both inputs are rejected, and only the difference between the inputs is amplified. Interference from 60 Hz power is the most frequently encountered common mode signal, and these circuits provide considerable screening of desired signals from unwanted noise.

3.3.4. Integrating and Averaging Amplifiers

Replacement of the feedback resistor in the inverting amplifier circuit of Fig. 3.11 yields the circuit shown in Fig. 3.14. Since the capacitor stores the charge fed back from the output, it acts as an integrator, and the time constant is a function of both resistor and capacitor values:

$$\Upsilon = 2\pi R_a C \qquad\qquad (Eq.\ 3.15)$$

The capacitor can be discharged and the integrator re-set by a shorting switch across C or by more complex circuit elements. If a feedback resistor is also employed parallel to the capacitor, the circuit acts as an averaging circuit, with a time constant of $2\pi RC$. This circuit acts as an inverting amplifier at low frequency and as an integrator at high frequency. It is a very practical circuit for eliminating high frequency noise from low frequency signals, in effect a low-pass active filter.

3.3.5. Other Op Amp Applications

The application of op amps is limited only by the imagination of the user, but those of us with little background may simply copy a tremendous variety of circuits from the various op amp "cookbooks" on the market. The

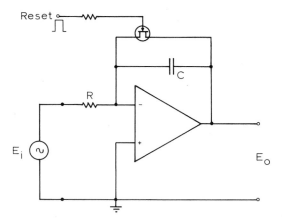

Fig. 3.14 An op amp in an integrator configuration. The output equals the input summed over time. The circuit has a time constant approximately equal to $2\pi RC$. The FET trigger shown will reset the output to zero with a pulse applied.

applications include comparators, limit detectors, current to voltage converters, sine wave generators, constant current sources, and many others. Op amps are available that can be run from battery current as low as ± 2 or 3 V or from single power supply voltages.

3.4. POWER SUPPLIES

For most commercial instruments, the power supply is either built in or provided by batteries, in which case the operation does not concern us. Many times, however, it will be convenient to put together a small power supply for a piece of "home-brewed" apparatus or simply to power a simple interface between one piece of equipment and another. The design of three different power supplies is described here, supplies which should suffice for most low-voltage applications.

3.4.1. Using the Zener Diode

The schematic in Fig. 3.15 shows a very simple power supply consisting of the following elements: The AC line is controlled by a switch (S), limited by a fuse (F), and brought to the primary winding of the transformer (T). An inexpensive neon indicator lamp is used to show when the power is on. The secondary winding is "center-tapped," making in effect two windings. The center tap is used as the ground reference point, and the ends of the two secondary half-windings are led through two diode rectifiers to convert the AC current to pulsating DC. The large filter capacitor (C) damps the

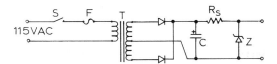

Fig. 3.15 A simple power supply with a Zener diode regulator (Z). S ≡ switch; F ≡ fuse; T ≡ transformer; C = filter capacitor (electrolytic type); and R_s ≡ current limiting resistor. This circuit will provide moderate regulation of the output with varying line voltage and load.

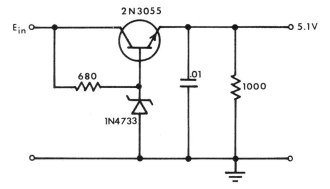

Fig. 3.16 An improved power supply circuit, with the Zener diode used to regulate the base voltage of a pass transistor. The purpose of this circuit is to provide a greater current handling capability while maintaining the voltage regulation with the Zener diode. The component values shown will provide up to 3 A with adequate heat sinking of the 2N3055 transistor.

pulsations to a small ripple, and this voltage is fed through a limiting resistor (R_S) and to the Zener diode (Z). The Zener diode is a device that passes almost no current (has high resistance) in the reverse direction up to a certain threshold, the Zener voltage, above which it passes current easily (has low resistance). It thus acts as a voltage regulator, and will limit the voltage fluctuations over a fairly wide range of load resistance (R_L). This circuit only provides limited voltage regulation, however, and cannot be used for very high currents. The values given in Fig. 3.15 will provide up to 0.2 A output with a 5.1 V Zener diode (such as a 1N4733).

3.4.2. Using a Pass Transistor

Better regulation and higher current output may be obtained with the circuit shown in Fig. 3.16, which employs the Zener diode to fix the base voltage of a pass transistor. With the values shown, this circuit will provide excellent regulation, and can supply currents of up to several amperes with an appropriate heat sink attached to the transistor. Caution: These transistors can reach soldering temperatures under short-circuit currents, so touch them

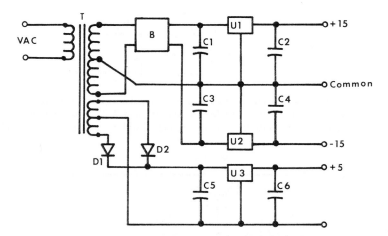

Fig. 3.17 A complete three-voltage laboratory power supply using solid state regulators of the LM78xx series. The following values will provide 1 A at 5 VDC and 150 mA at ± 15 VDC:

U1	LM7815CT voltage regulator
U2	LM7915CT negative voltage regulator
U3	LM340T-5 voltage regulator
B	Bridge rectifier, 100 PIV
C1, C3	1000 μF, 50 V electrolytic capacitor
C5	4700 μF, 10 V electrolytic capacitor
C2, C4, C6	0.01 μF, 100 V disc ceramic capacitor
D1, D2	1N4002 diode
T	MPC-Y-15 transformer, Signal Trans. Co.

carefully. The lack of overload protection is a serious disadvantage of this circuit.

3.4.3. Using Integrated Circuit Regulators

The simplest method for providing regulated voltages is to take advantage of the integrated circuit voltage regulators now on the market for about $1.50. The schematic of Fig. 3.17 shows a three-output power supply suitable for powering a wide variety of both digital and analog circuits in the laboratory. The transformer listed has two secondaries, one of which is used for the ± 12 or 15 V supplies and the other for the + 5 V supply. This supply is capable of providing 1 A at 5 V and about 150 mA each from the ± 12 or 15 V supplies. The 7805 regulator must have a good heat sink attached to it, preferably with some heat transfer paste applied to the contact surface (a zinc oxide sunburn ointment will do). The circuit is internally protected against overload, however, and will not burn out with the output short-circuited.

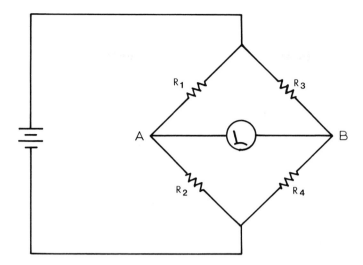

Fig. 3.18 The Wheatstone bridge. The current through the meter shown in the center is zero when $R_1R_4 = R_2R_3$. The bridge may be used in various ways for measurement of resistance based signals (see text).

3.4.4. Modular Power Supplies

Several manufacturers now offer complete modular power supplies in a variety of output voltage and current combinations. A typical unit will be encapsulated in a $2.5 \times 5 \times 10$ cm package, and may have output capabilities of ± 15 V at 150 mA each, and $+ 5$ V at 1 A, with only a few mV voltage "ripples" at the highest current loads. These units are convenient but considerably more expensive than the simple supplies given above. The choice is a matter of time, supply requirements, and cost.

3.5. THE WHEATSTONE BRIDGE

A simple but enormously useful circuit for physiological measurements of many kinds is the *Wheatstone bridge*, shown in Fig. 3.18. This circuit, originally described for resistance measurement by Christie in 1833 (and later made known by Wheatstone), consists of four resistance elements, R_1–R_4, arranged in such a way that if $R_1/R_2 = R_3/R_4$, then the potential between the output points A and B is zero and the bridge is said to be balanced. For resistance measurement, one of the fixed arms, say R_1, is replaced by the unknown resistor, and the bridge is balanced by a known adjustable resistor at one of the other positions. The values of the three known arms are then used in the ratio calculation to obtain the unknown arm's value. This circuit is rarely used directly for resistance measurement but is used in a wide variety of transducer applications. To list only one, the circuit of Fig. 3.18 could be used as a thermistor thermometer by replacing one of the arms of the bridge with a thermistor. A *thermistor* is a resistor constructed of special materials that give it a high negative temperature coefficient, i.e., as the temperature rises, the resistance falls. The temperature of the thermistor could be measured either by balancing the bridge and comparing the computed resistance with the temperature–resistance curve supplied with the thermistor or by measuring the unbalanced current or voltage. The output of a single thermistor will give a non-linear curve with temperature, but two matched thermistors may be employed in a bridge circuit to produce a resistance that is highly linear over useful ranges of temperature (Fenwal Bulletin L-9A).

3.6. ELECTRICAL MEASUREMENTS: VOLTAGE, CURRENT, RESISTANCE

3.6.1. The General-Purpose Voltmeter

The most common and generally useful electrical measurement device is the hand-held or portable voltmeter. This can be constructed with a moving-coil type of panel meter, but the newer types are increasingly making use of a digital display, with liquid crystal displays (LCDs) becoming more common (Fig. 3.19). Voltmeters are generally equipped with two measuring leads, or probes, one black for the ground or relatively more negative connection and the other red for the positive connection. Most instruments are equipped with range switches to allow measurement of voltages in several ranges, both AC and DC. The input impedance of the moving-coil meter types is usually fairly low, unless a high input impedance amplifier circuit is included, limiting their usefulness in measuring many signal types. The newer LCD meters, however, often have very high input impedances, high enough to be used directly for measurement of cell potentials, or even with many high-resistance electrodes having low current source ability. These meters may be purchased for under $100 and should not be overlooked as convenient and versatile measurement devices.

As well as making voltage measurements, most of these meters also have provisions for making current and resistance measurements. When making resistance measurements, the two test leads have a voltage applied across them internally, and the instrument is usually constructed to measure the current flowing through the resistive element under test. This feature should be used with caution, since some of the older meters, in particular, apply voltages sufficient to burn out sensitive circuit elements. Some of the newer meters allow selection of the test current for resistance measurement, either

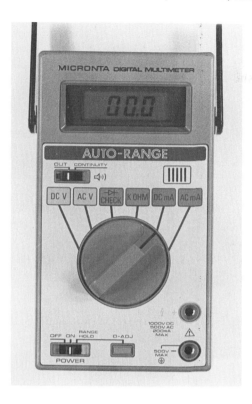

Fig. 3.19 A hand-held digital multimeter. This particular model has provisions for measuring AC and DC voltage and current, resistance, diode integrity, and continuity. The LCD display allows the use of low-current circuitry, giving a long battery life. The input impedance is high enough so that it can be used directly as a measurement device in many laboratory situations.

in high and low ranges or in steps. The resistance measuring function is also very useful in tracing circuit paths, since open paths show infinite resistance.

3.6.2. The Capacitance Meter

The measurement of capacitance is not directly required in physiological studies, but it is often handy to be able to measure the value of circuit components. Many types of capacitors have wide tolerances in manufacture, and the marking systems in common use are often mysterious. One type of capacitance meter is shown in Fig. 3.20. This meter works by taking advantage of the charging characteristics of the *RC* circuits discussed in Section 3.2.3. Starting with a fully discharged capacitor, the instrument

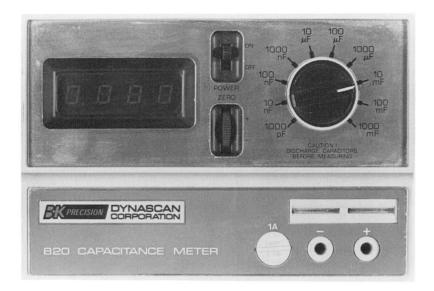

Fig. 3.20 A capacitance meter. This convenient hand-held model relies on the charge timing for measurement (see text).

measures the time required to reach a fixed potential (voltage) with a known resistor in series. The time is directly proportional to the capacitance value, which is displayed in digital form.

3.7. ELECTRICAL RECORDING

The circuits discussed so far apply only to the processing and measurement of electrical signals, but much of the time the electrical information must be recorded in some fashion, if not for immediate visualization, then for later retrieval and analysis. The forms of electrical signals required for display, such as deflection of a meter coil or signals to a digital display, are usually not those that can be recorded, so an additional transduction step or two is required. The nature of the transduction needed is a function of the kind of information that is to be stored and the form in which it will be stored.

3.7.1. Strip-Chart Recorders

The most familiar type of recording device is probably the strip-chart recorder, the latter-day descendant of the smoked drum kymograph. Most strip-chart recorders are of the servomechanism type, a functional block

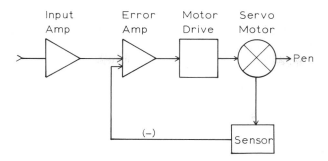

Fig. 3.21 Functional block diagram of a servo recorder. The error amplifier compares the input with the output of the pen position sensor. If they differ, an error signal of opposite sign causes the motor drive to re-position the pen, until the error signal is zero.

diagram of which is given in Fig. 3.21. The details of each functional block may differ from one model to another, but the principle of applying a negative feedback correction to a device which sums the correction and the input signal to control the mechanical location of the pen on paper is fairly universal. Commercially available recorders offer a wide range of input scales, chart speeds, channel numbers, and options for pen lift, timed operation, etc.

One common feature of these recorders is that their *frequency response* is limited. The model shown in Fig. 3.22, for example, has a full scale frequency response of less than 1 Hz, and about 2 Hz at only 10% of full scale. For more rapidly varying signals, oscillographic recorders of various manufacture are available, such as the one shown in Fig. 3.23. The design principles of these recorders are similar to those of the strip-chart types, but the emphasis in design is on providing a low mass, and therefore low acceleration resistance, of the pen and pen positioning mechanism. These recorders are frequently seen in multi-channel design and are employed for recording a wide variety of physiological signals, including blood pressure and EKG traces. For even higher frequencies, when a real-time recording is desired, very fast recorders are now coming on the market that make use of ink-jet technology borrowed from computer printers. Since these recorders generally have no moving mechanical parts, they avoid the simple problems of inertia and acceleration inherent in the other types.

Most commercial recorders have provisions for accommodating a range of input signals, and usually have a selector switch that allows adjustment of the full scale from zero to various limits. Frequently, however, one wants to record a small signal variation occurring superimposed upon a large base potential. For example, one might want to see changes of a few millivolts around a constant voltage of 3 V. Although a few recorders include *zero suppression* circuitry, it is frequently of limited range, and the circuit shown

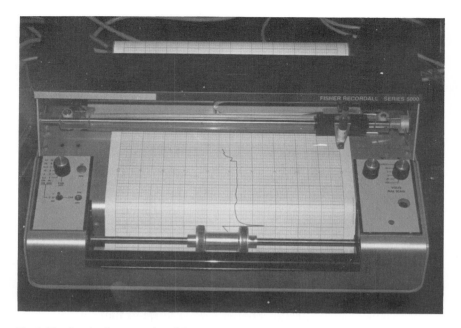

Fig. 3.22   A strip-chart recorder of the servo type, manufactured by Houston Instruments. The position sensor is a capacitor formed by an aluminized belt connected to the pen. These recorders are relatively low in cost but have a slow response to changes in the input signal.

in Fig. 3.24, which can be made for a few dollars' worth of parts from any electronics supply house, is quite useful. A large mercury battery provides a very stable reference voltage, and by selecting resistor values, almost any baseline value may be subtracted from the input signal. The use of a precision 10-turn potentiometer and 10-turn dial (e.g., Clarostat 73JA and 411) will provide calibrated, continuously adjustable zero suppression.

3.7.2 Tape Recorders

High frequency and other complex AC electrical signals may be directly recorded on a magnetic tape recorder, provided that the input voltage to the recorder is scaled in such a way that the tape is not saturated. Slower signals and DC voltages cannot be successfully recorded on magnetic tape in analog form but may be stored on tape in digital form (see Section 3.9). The advantages of tape storage are that very long records may be stored with a minimum of cost and space, and the record may be replayed at a later date for display or analysis. The disadvantages are that tape is a somewhat fragile medium and that finding a segment of a long record can be a frustrating chore unless a careful catalog system is kept up.

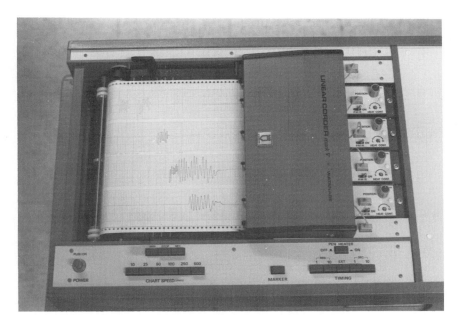

Fig. 3.23 An oscillographic recorder manufactured by Watanabe. Although the principle is similar to that of the servo pen recorders (Figs. 3.21 and 3.22), the pen positioning and sensing mechanisms are designed for low inertia and high response velocity. Signals of up to about 100 Hz can be faithfully recorded.

3.8. THE OSCILLOSCOPE

The heart of the oscilloscope is a cathode ray tube (CRT) equipped with electrostatic deflection circuitry for the electron beam. The beam appears as a dot on the screen, and a wide range of user controls are provided to govern the movement of that dot in such a way as to allow visualization of both low- and high-frequency voltage patterns. By convention, positive voltages applied to the Y-input produce a deflection of the beam toward the top of the screen, negative voltages toward the bottom. Since the CRT itself requires about 50 V per inch of deflection, extensive circuitry is required internally to provide amplification and linearization of the deflection patterns, and usually a wide choice of amplification is provided by the front panel controls. The X-input is most commonly driven by a time sweep generator. That is, the horizontal axis is treated as a time axis, with increasing time toward the right. The beam proceeds from left to right, reappearing at the left edge as soon as it leaves the right. The time sweep may be driven at a variety of rates, so that 1 cm on the screen reticle may represent various units of time.

Most oscilloscopes also provide for a "triggered" sweep and allow a choice

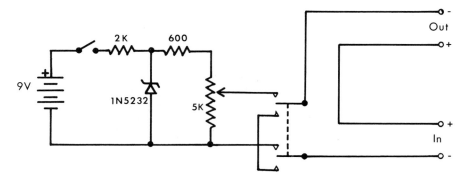

Fig. 3.24 A circuit for zero suppression in conjunction with signal recording. The 1N5232 is a 5.6 V Zener diode, so the 5K variable resistor gives a 0–5 V zero suppression signal. Replacement of the Zener diode with a precision reference (such as that of Analog Devices AD584K) gives a more precise setting, and when used with a 10-turn potentiometer and dial (such as Clarostat 73JA and 411), provides accurate and reproducible zero suppression.

of trigger parameters. This means that the time sweep of the beam from left to right begins when the conditions of the sweep trigger have been met, and the slope of the input signal, its magnitude, or sometimes other features may be used to trigger the sweep. For repetitive events with varying time delays between them, this means that the time base of the visible trace always begins at about the same point in the event, which makes interpretation of the viewed data much easier.

Besides the normal Y/T mode, many oscilloscopes may also be driven in an X/Y mode, wherein both the horizontal and vertical deflections are driven by input signals. This is a particularly useful mode in visualizing phase relationships between different frequency signals. Finally, a "Z" input is offered on some oscilloscopes, which either turns the beam on and off or modulates its intensity. This mode is generally used in electronics testing and is not very important in physiological applications.

There is a wide variety of oscilloscopes on the market, ranging in price from a few hundred dollars to several thousand, depending on the features offered. One common type is shown in Fig. 3.25, a dual-beam model. The dual-beam oscilloscope and the "dual trace" scope are not the same. The former has two complete, independent electron guns, whereas the latter employs a single beam with alternation or "chopping' between channels to display two traces. Various storage modes are also offered on the more elaborate instruments, allowing recall of events of interest. Traces from an oscilloscope may be recorded with a variety of cameras, using either sheet or roll film.

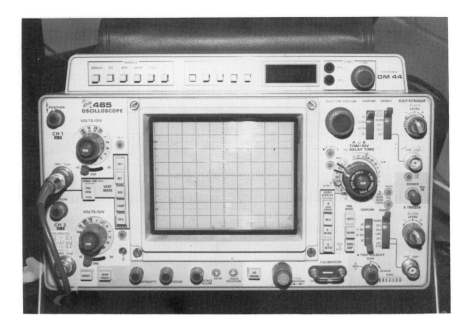

Fig. 3.25   A Tektronix dual-beam oscilloscope. The oscilloscope is an indispensable aid in testing and trouble shooting, as well as in visual recording of rapidly varying signals.

3.9.   DIGITAL TECHNIQUES

3.9.1.   What are Digital Data?

Until now, the discussion has dealt exclusively with electrical signals that continuously vary in magnitude, or *analog* signals. We, of course, characterize these voltages or currents in digital (decimal) form when we write them down. In order to translate these voltages into a form which can be dealt with by computers, however, the analog signals must be represented by a binary number. The binary representation of the analog signal must also be limited in length, since the computers and associated circuitry commonly in use have limitations on the length of the numbers they can work with.

Let us imagine that we have only an 8-bit number length as our maximum for a particular computer. This means that the largest number we can represent is (binary) 1111 1111, or (decimal) 255, and a total of 256 different numbers are available (including 0). Let us also imagine that we have a signal whose strength varies between 0 and 10 V. If we set the maximum range equal to 255 arbitrary number units, then each digit increase will represent a change of 39 mV, which is the *resolution* of our binary digitizing system. Similarly, 12-bit binary digitizing would provide 4096 steps, with a 2.4 mV resolution, and 16-bit digitizing would provide 65,536 steps with 0.15 mV resolution. The choice of resolution for any given application depends, then, on the resolution required and the data-handling capabilities of the computer (or other digital acquisition device) available.

Most of the microcomputers that are gaining popularity in laboratories are based on 8-bit *bytes*, but there are various means of increasing the resolution in spite of this limitation. A 12-bit digitizing circuit for 8-bit computers that can be built for under $100 has been described recently (Cameron, 1983) which works by splitting the 12-bit number into two pieces and putting the pieces together under program control to reconstruct the full 12-bit number.

Three further considerations are important in choosing or designing a digitizing system: speed, memory capacity, and cost. Speed is not as much of a limitation as it once was, since new circuits appear almost every day for analog-to-digital conversion (ADC), and the current interest in digitizing consumer audio and video equipment has led to the development of very fast, very high resolution circuits. The cost does become a factor, since an 8-bit conversion chip may be purchased for a few dollars, 12-bit chips for $10 to $50 or more, and 16-bit chips or modules for around $200 and up. Within each resolution category, there is a direct relationship between speed and cost, too.

3.9.2. An Example Digitizing Problem

The problem of conversion speed very quickly becomes entangled with the final problem of memory capacity. Let us take a practical example to illustrate the problems in designing a digitizing system. Let us say that we have a device that measures the blood pressure pattern in a monkey, and that we want to design a digital data acquisition system. The monkey's pulse rate is 120 per minute, or 2/sec, but some higher frequency components of the pressure wave make it desirable to faithfully represent signals up to about 20 Hz. In order to faithfully reproduce a 20 Hz signal with digital sampling, we need to sample at about 10 times that frequency, or 200 Hz. Let us say that we also need a resolution of only 8 bits, so that our *data rate* will be 1600 bits, or 200 bytes sec^{-1}, and the maximum conversion time allowed will be no more than 5 msec. At this data rate, the typical microcomputer with 64K (i.e., 65,536 bytes) of memory capacity will be full in less than 6 minutes. Assuming we can arrange the transfer of the data from the computer's memory to a floppy disk without losing any data, we can extend our maximum measurement time to perhaps 30 min, depending on the particular disk system's capacity. That may or may not be adequate for the intent of the experiment, but gives some idea of the constraints imposed.

The problem is much worse if greater resolution is required or higher frequency components are present. The same problem with a hummingbird, which has a pulse rate of perhaps 300 min^{-1}, highest frequency of the pressure wave at 50 Hz, and 12-bit resolution required (using 2 bytes per 12-bit conversion), would produce a data rate of about 1 Kbyte sec^{-1} and would reduce the time needed to fill memory to about 1 min. The conversion time would only be 1 msec, plus some program time, which gets us into the realm of expensive circuits, and machine language programming requirements.

There are ways to reduce the problems posed above, one of which is to sample the signal of interest at intervals, rather than making continuous measurements. For example, it might be sufficient to record the monkey's blood pressure twice each hour for 1 min, or for 5 sec each minute, reducing the data storage requirement by several orders of magnitude. There are also some very sophisticated techniques for varying the sampling rate according to the rate of change of the signal frequency; these techniques have reached a state of high refinement in the area of speech analysis and synthesis, but are probably beyond the needs and capabilities of most physiologists.

3.9.3. When to Digitize?

The primary consideration in deciding whether or not to digitize (computerize) any particular data-gathering activity is probably the time required to design the system and get it operating smoothly vs. the time required to do it in an alternate way. Even with commercially available data acquisition modules and their associated software (i.e., the programs necessary to run the hardware), the time required to convert an experiment to computerized data acquisition can be very considerable. If the data acquisition circuitry is to be custom built, then the time required will be even greater. It is certainly true that once running, these systems can save enormous amounts of time in routine data processing. There are even some kinds of experiments that we would not attempt at all without automated data acquisition and processing. The initial time investment, however, should probably be analyzed in the same terms as any capital investment: will it yield sufficient profit over a long enough period of time to repay the initial investment? In most people's experience, the answer is negative unless the experiments are likely to be repeated often with little modification.

Cost is also a consideration in whether to automate data acquisition, but not as great a factor as it was even a few years ago. A complete microcomputer system, including two disk drives and a printer, may be purchased for under $3000, which is no more than average for a piece of laboratory equipment. The data acquisition circuitry may be either purchased or built for a few hundred dollars more. Time again is probably the major cost and is difficult to estimate accurately, especially for the programming effort required to make the data acquisition system useful.

There are further points to consider in relation to programming. One is that although the manufacturers of data acquisition equipment may provide some software to control the equipment and manage data storage in some reasonable format, they cannot know what you want to do with it once it is stored. Programs therefore must be written to perform whatever analysis is required for the experiment at hand. Research, by its very nature, dictates that what is required is probably not a common application, so the software often must be written from scratch. The second problem is that extensive programming requires time and skills that the investigator may not have, so the time and cost analysis must include the necessary learning time.

3.9.4. Digital Storage Media

The most common and practical digital storage medium nowadays is the *floppy disk*, a descendant of the magnetic tape. The disk is made from a plastic and coated with a magnetic material. A read/write head is brought into close proximity to the coated surface of the spinning disk (by various mechanical means), and the data are recorded in concentric circular tracks. The numbers of tracks, density of recorded information, size, and so forth vary considerably, but $8''$, $5\frac{1}{4}''$, and now 3 to $3\frac{1}{2}''$ disks are common, and the achievable densities are continually rising.

For much greater storage capacity, the *hard disk*, or Winchester disk, provides higher bit densities and much closer track spacing, due to the greater mechanical stability of the medium. These devices until recently were confined to large computer systems, but the prices have fallen rapidly and are now in a reasonable range for microcomputer use. The capacities of hard disks for microcomputer systems range up to about 20 megabytes, as compared with a few hundred kilobytes for floppy disks.

Tape recorders may also be used for storage of digital data, with the data recorded in various coding schemes on tape varying from $\frac{1}{4}''$ to $\frac{1}{2}''$ in width. The most popular models are usually seven to nine track, so that a number of different signals may be simultaneously recorded, and a large tape reel may provide several hours of data capacity. The tape systems tend to be more fragile than disks, however, and do require a careful catalog system.

LITERATURE CITED

Cameron, J. N. 1983. A high-resolution analog-to-digital converter for the TRS-80. Byte 83: 378.
Fenwal Electronics. Linear Thermistor Networks.

SUGGESTED FURTHER READING

The Radio Amateur's Handbook. The American Radio Relay League. (There is a new edition of this each year, which contains excellent discussions of many aspects of electronics.)
Linear Applications. National Semiconductor, Inc. Vols. I & II. (Contains many useful circuit applications with low-cost components.)
Wobschall, D. 1979. Circuit design for electronic instrumentation; analog and digital devices from sensor to display. McGraw-Hill, New York. A text oriented toward applications of electronics to scientific measurement and instrumentation. Not overly technical, and contains many useful circuits.)

CHAPTER

4

OXYGEN MEASUREMENT

4.1. BASIC PROPERTIES

Atomic number:	8
Atomic weight:	15.9993
Valences:	$+2$
Melting point:	$-218.4°C$
Boiling point:	$-182.96°C$
Density, 0°C:	1.429 g L^{-1}
Normal form:	O_2

Miscellaneous. Positively paramagnetic; natural isotopes ^{18}O (stable, 0.2%) and ^{17}O (stable, 0.04%); makes up 49.2% of the earth's crust by weight; highly reactive; prepared from liquefaction of air, electrolysis of water.

4.2. HISTORY

Oxygen was discovered in 1774 by Joseph Priestley, who prepared the gas from mercuric oxide by heating and observed that it supported combustion. He gave it the name "dephlogisticated air," which was replaced with "oxygéne" by French chemists in 1787. The atomic weight of naturally occurring oxygen was once taken as the standard and assigned a value of 16.000, but the single isotope ^{12}C is now assigned the standard value of 12.000, and the atomic weight of oxygen and other elements reflects the natural abundance of different isotopes.

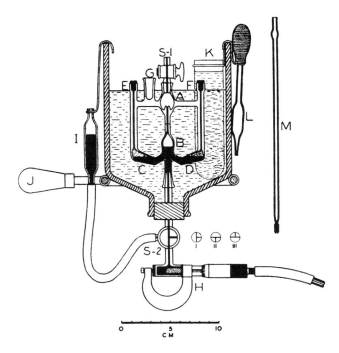

Fig. 4.1 The Scholander micro gas analyzer. The main components are the extraction chamber (A and B), the two side arms for reagents (C and D), and the micrometer syringe (H) for adjusting the volume at various stages of analysis to the mark between A and B. Operation is discussed in the text, and full details are given in Scholander (1942, 1947). (Reprinted with permission from J. Biol. Chem.)

4.3. OXYGEN MEASUREMENT IN GASES

4.3.1. Manometric Methods

Although earlier methods and apparatus were described, the standard method for many years for the analysis of oxygen in gas samples was the Scholander apparatus. This apparatus, shown in Fig. 4.1, is a descendant of the Van Slyke apparatus (Section 4.4.2) and various other early manometric devices. It consists of two small chambers (A and B), one of which is used to contain the gas sample and the other to compensate for thermal variations in the gas volume. By tilting the apparatus and then shaking it, first a CO_2 absorbent and then an oxygen absorbent solution may be introduced from the side arms (C and D). The micrometer burette is used to position a mercury drop at an index mark between the two chambers. After each addition, the reduction in volume is measured by re-adjusting the micrometer burette, and the percentage composition of the sample is then calculated by simple ratios.

Detailed instructions for the construction, assembly and the use of this apparatus are given by Scholander (1947), as well a reference to an earlier, simpler version (Scholander, 1942).

4.3.2. Analysis Based on Paramagnetism

Although most gases are slightly diamagnetic, meaning that they are repelled out of a magnetic field, oxygen, and to a lesser extent nitric oxide (NO) and nitrogen dioxide (NO_2), are paramagnetic. For practical analysis, oxygen is the only significant paramagnetic gas, since the oxides of nitrogen are only trace components, and their magnetic susceptibilities are considerably less than that of oxygen. The Beckman paramagnetic oxygen analyzer takes advantage of this property, as shown in the accompanying diagram (Fig. 4.2). A dumbbell of hollow glass spheres is suspended in a magnetic field by a quartz fiber in such a way that the magnetic attraction tends to rotate the dumbbell toward the magnet poles, and the torsion on the fiber opposes this force. When a gas containing oxygen is introduced around this assembly, the magnetic susceptibility of the oxygen causes a change in the magnetic flux density between the pole pieces and leads to a change in the balance point of the fiber/dumbbell assembly. By mounting a mirror on the fiber and bouncing a beam of light off it, the dumbbell rotation may be indicated on a calibrated scale. The method is quite sensitive, providing resolution of $\pm 0.05\%$ for a 5% oxygen full scale. This instrument has for many years been a standard method for measurement of changes in respiratory oxygen, both in animal and plant studies. It may be used in both flow-through and single sample injection modes.

The magnetic susceptibility of oxygen is strongly temperature dependent, a property which has led to another type of analysis instrument. The Hays Magno-Therm employed a combination of paramagnetism and thermal conductivity (see the next section). The sensor was a heated wire whose resistance was a function of temperature, thus lending itself to application in a Wheatstone bridge circuit (see Section 3.5). The sensor was placed in a dead-end channel perpendicular to the gas flow, and a strong magnetic field was applied across it. The magnetic field drew relatively cool oxygen from the gas stream, which was then heated by the hot-wire sensor. As it was heated, it lost its paramagnetism and so was pushed out of the field by the cooler, more magnetically susceptible oxygen. This process set up a gas flow around the hot-wire sensor, cooling it and changing its resistance in an easily measurable way. Correction for the effects of gas flow was accomplished by employing a second identical cell without the magnetic field as another arm of the bridge. The sensitivity of this method was somewhat less, $\pm 0.25\%$ up to 20% oxygen and $\pm 2.5\%$ up to 100% oxygen, but the

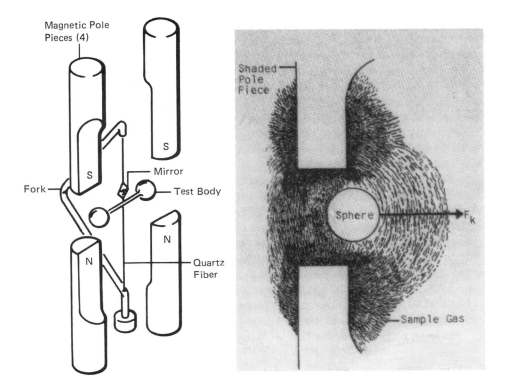

Fig. 4.2 Schematic diagrams of the operation of the Beckman paramagnetic oxygen analyzer. On the left, the physical arrangement of the detector is shown. It consists of a dumbbell-shaped object suspended between the pole pieces of a magnet by a thin quartz fiber. Since oxygen is attracted to the most intense area of the magnetic field, the dumbbell is displaced as shown at the right. The displacement can be accurately measured by bouncing a light beam off the mirror attached to the suspending fiber. Magnetic susceptibility: sphere, k_0; sample gas, k. Note: displacement force (F_k) increases as percentage of oxygen in sample gas increases. (Courtesy Beckman Instruments, Inc.)

device was more rugged than the paramagnetic analyzer and easier to adapt to a wide range of concentrations.

4.3.3. Gas Chromatography

The principle of separation in gas chromatography is differential retention of gases in a stationary column material, through which a mobile (gas) phase carries the sample to be identified and measured. For the atmospheric gases, a basic chromatograph might appear like the schematic in Fig. 4.3. There

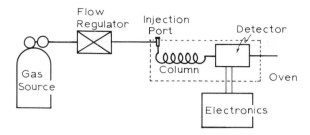

Fig. 4.3 A functional block diagram of a simple gas chromatograph. See text for discussion.

is first a flow regulation system, which usually consists of a helium cylinder with pressure and flow regulation. The gas phase is generally brought through the reference side of the detector, past an injection port for introduction of the sample, then through a column containing the stationary phase, and finally through the sample side of the detector.

A variety of column materials are suitable for separating O_2 from the other atmospheric gases. One good material is Poropak Q, which retains gases differentially based upon molecular size and a "molecular sieving" effect. In practice, a metal tubing column with a 1 to 6 mm inside diameter and 1 to 2 m long is packed with this material and maintained at an elevated temperature to speed and increase separation.

There are also a variety of detectors in use in gas chromatographs, but the simplest and most suitable for the atmospheric gases is the *thermal conductivity* detector. This detection method takes advantage of the large differences in thermal conductivity among the atmospheric gases and dictates the choice of He as the mobile phase, or carrier gas, since it has the highest thermal conductivity of all the common gases (Table 4.1). The detector is arranged as shown in Fig. 4.4, with two thermistors or other temperature sensors, one mounted in the reference stream, the other in the sample stream. The two sensors may be connected as two arms of a Wheatstone bridge (see Section 3.5), with sufficient current flowing through the bridge to heat the thermistors well above ambient temperature. At a steady flow of He, the thermistors will reach a steady state in which the rate of heat gain from self-heating is just balanced by the rate of cooling from the He stream passing over them. When a quantity of another gas passes through the sample side mixed in the He stream, the thermal conductivity is reduced from that of pure He and the thermistor heats up, causing its resistance relative to the reference side to drop. This induces an imbalance in the bridge circuit that may be measured with great sensitivity.

The gas chromatograph produces an electrical signal pattern like that shown in Fig. 4.5. The peaks for each gas are not symmetrical, and the

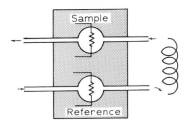

Fig. 4.4 A schematic diagram of a thermal conductivity detector for gas chromatography. The sample and reference elements are often wired as arms of a Wheatstone bridge circuit.

TABLE 4.1
Thermal Conductivity of Some Common Gases[a,b]

Gas	T.C.	Gas	T.C.
H_2	416.0	N_2	58.0
He	352.0	CO	55.9
Ne	108.7	NH_3	52.2
CH_4	72.1	A	39.7
O_2	58.5	CO_2	34.0
Air	58.0	SO_2	20.6

[a]Reproduced with permission from Radford (1964).
[b]The units of thermal conductivity are $^{\circ}cal\ cm^{-1}\ sec^{-1}\ ^{\circ}K^{-1}$.

amount of gas in the sample is not always linearly related to the peak height. More often the peak area must be measured, either manually or by electrical integration. Gas chromatographs suitable for atmospheric gas analysis are relatively simple in their design requirements, but a great many extra features may be added by the manufacturer to make the operation more automatic. The method is accurate down to ppm concentrations of many gases, depending on the values chosen for gas flow, column length, temperature, etc.

4.3.4. Mass Spectrometry

Although the design of mass spectrometers for actual applications is a complex field, the principles of their operation are not difficult to understand. The operation may best be understood with the aid of the schematic diagram of Fig. 4.6, considering each major element in order.

The Inlet.   Since the mass spectrometer operates with a minute quantity

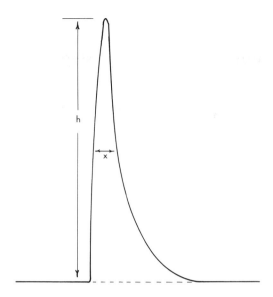

Fig. 4.5 An idealized gas chromatogram peak, showing the asymmetry on both the rising and falling edges. Multiplication of the peak height (h) by the width at half height (x) will give a fair estimate of the peak area so long as the tailing (shown toward the right) of the peak is not too severe.

of sample in a high vacuum, the inlet system must be designed to admit a carefully controlled amount of the gaseous material. In the process, no fractionation of the gas must be allowed to occur, which usually means that viscous flow is required, rather than "molecular flow," or diffusion. For a discrete sample, one way to do this is to simply expand a small amount of

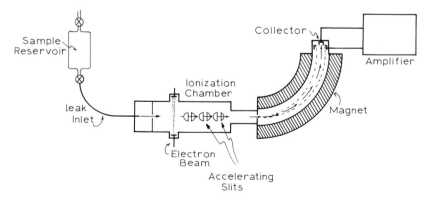

Fig. 4.6 Functional block diagram of a mass spectrometer. The function of each section is described in the text.

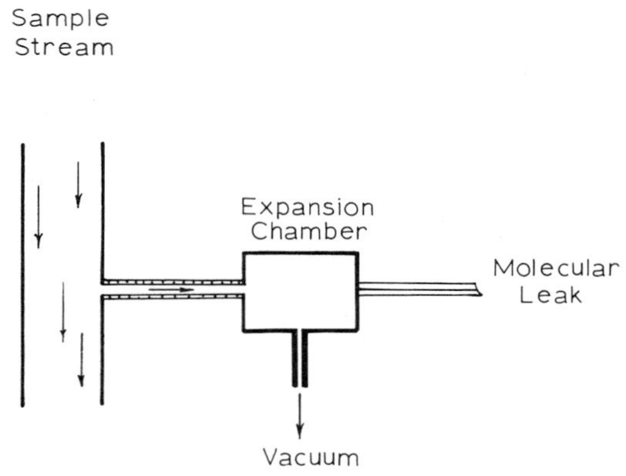

Sample
Stream

Fig. 4.7   Schematic diagram of a two-stage pressure reduction system for a mass spectrometer inlet. This arrangement is useful for sampling continuous gas streams.

gas into a large evacuated container, from which a small quantity may be allowed to flow under a reduced pressure gradient. The physical device used for the inlet may be simply a fritted disc, a fine bore capillary, or a hole in a metal foil. For sampling a continuous stream, it may be necessary to arrange a two-step pressure reduction, allowing gas flow from the intermediate pressure stage, as shown in Fig. 4.7. For most applications the inlet flow must be constant, so provisions for monitoring and controlling the inlet pressure are designed into the instrument.

The Accelerating Chamber.   The molecules of gas enter the accelerating (or ionization) chamber, which is maintained at a high vacuum and relatively high temperature, perhaps 200°C. As the molecules enter, they cross through an electron beam, which usually consists of a hot carbonized tungsten filament to generate electrons and two positively charged slits to accelerate the electrons and draw them off. Collisions between the entering neutral molecules of gas and the electron beam produce negatively charged ions, which are then focused and accelerated into the next section by an electric field. The velocity with which the ions enter the next section depends on the electric field strength, which in most instruments may be varied between about 5 and 100 V.

The Analyzer Tube.   The accelerated beam of ions enters a tube, which is bent as it passes through a large magnet, as shown in Fig. 4.6. The ions are deflected from their straight line of flight with a trajectory whose radius of curvature (r) depends on the mass (m) of the ion, its charge (e), its velocity (v), and the applied field strength (H) according to the following equation:

$$r = (2Vm/H^2e)^{0.5}$$

(Eq. 4.1)

Thus, for a particular ion mass and charge, the voltage and field strength may be adjusted so that the radius of curvature just matches the radius of curvature of the analyzer tube. Ions of other mass and charge collide with the walls, where they are neutralized and pumped away by the vacuum system. The choice of accelerating voltage and field strength depends on the particular instrument and the intended application. For steady measurement of a particular ion species, both settings may be held constant, whereas for measurement of a mass range, either the voltage or the field strength may be swept over an appropriate range of values. Some instruments have a provision for sweeping only one of these parameters, e.g., with variable voltage and fixed permanent magnets.

The Detector. The simplest kind of detector for the very weak ion current (roughly 10^{-18} to 10^{-10} A) is a metal cylinder employed as the grid of an electrometer tube, which provides a high impedance, high-gain amplification. For fast sweep instruments, various electron multiplier devices are in use. In many instruments, more than one detector may be provided, so that two masses close to one another may be measured simultaneously. This is particularly useful in stable isotope work, where an example would be the measurement of the ratio between $^{12}CO_2$ and $^{13}CO_2$, masses 44 and 45, or $^{16}O_2$ and $^{18}O_2$, masses 32 and 36. (Mass 34, $^{16}O^{18}O$, is actually the commoner molecule.)

Application to Physiology. The most powerful application of the mass spectrometer in physiology is continuous measurement of the principal respiratory gases. For example, oxygen (mass 32), CO_2 (mass 44), and nitrogen (mass 28) may all be monitored simultaneously by rapidly sweeping the mass range, and it is no more trouble to add in measurements of some anesthetics and single-breath stable isotope gases. There are some design compromises to be made, of course. To sweep rapidly enough to follow changes during the breathing cycle, the beam current must be higher, which means a higher sampling flow through the leak. A principal disadvantage of the method is the size, complexity, and expense of the apparatus, although several instruments designed especially for physiological studies have appeared in the past several years.

4.3.5. Galvanic Methods

The principle of galvanic analysis is that the material of interest (in this case, oxygen) may be made a reactant at one of a pair of electrodes, without an externally imposed voltage, thereby generating an electrical current. The application of this principle to O_2 analysis was first described in 1933, and since then a variety of cells for various applications have been described. One possible cell consists of a silver cathode, KOH electrolyte, and cadmium anode, in which the cathode reaction is:

$$O_2 + 2 H_2O + 4 e^- = 4 OH^- \qquad \text{(Eq. 4.2)}$$

and the anode reaction is:

$$2 Cd = 2 Cd^{++} + 4e^- \qquad \text{(Eq. 4.3)}$$

Porous carbon and platinum are alternate cathode materials, lead an alternate anode material. This type of analyzer can be constructed in a variety of configurations, for both discrete sample and continuous stream analysis, and may provide extreme sensitivity, down to part per million levels of O_2 (Hersch, 1964). From time to time commercial instruments based on the galvanic principle have appeared, but I am aware of none at present that are suitable for physiological measurement. Information given by Hersch would suffice for construction of an apparatus for special purposes, though one must be careful not to introduce other reactive materials which may "poison" the cell.

4.3.6. Oxygen Electrodes

Although the use of oxygen electrodes for measurements in gases is quite feasible, the method is better adapted to the measurement of dissolved oxygen in solution, so a complete discussion is deferred to Section 4.4.3. Slight water permeability of some membranes in common use makes drying of the electrolyte a particular problem for gas measurements. Another possible pitfall is that a rapid flow of gas over the electrode face may cause evaporative cooling, thus shifting the calibration of the electrode.

4.4. OXYGEN MEASUREMENT IN AQUEOUS SOLUTION

4.4.1. The Winkler Method

The Winkler method for measurement of dissolved oxygen in water should now be of only historic interest in physiology, but since it is still frequently encountered, a short discussion follows. The method depends on a series of reactions, usually carried out in a specially designed bottle that allows serial introduction of reagents without exposing much of the sample to the atmosphere. First, a manganous hydroxide precipitate is dispersed in the sample, which becomes oxidized to higher valence states by dissolved O_2 in the sample. Subsequent addition of iodide and acid causes the reduction of the manganese to its original divalent state, with an equivalent liberation of free iodine. The iodine, now present in the sample exactly in proportion to the original dissolved oxygen, is titrated with a thiosulfate solution.

The advantage of this method is that it is cheap and easy to carry out, even under field conditions, requiring only simple chemicals and some glassware. The disadvantages are considerable, however. Many different substances interfere with the reactions at various stages, and may lead to either an increase or a decrease in the apparent dissolved oxygen value obtained. Unless the concentrations and effects of all interfering substances are known in advance, the method yields uncertain data. There is also a consistent problem of individual operator differences, which make it difficult, if not impossible, to compare data from different laboratories or even from different investigators in the same laboratory. In view of these many problems and the widespread availability of superior methods, the Winkler method should probably not be used in physiological studies.

4.4.2. Manometric Methods

Manometric methods and apparatus, especially for blood gas analysis, have been described in number practically since the discovery of oxygen in the 18th century (see also Chapter 10). The most enduring of these is undoubtedly the apparatus of Van Slyke, originally described in 1917 and still sold today in modified form. The apparatus is shown in Fig. 4.8. The principal elements are an extraction chamber and a mercury barometer. To analyze a sample containing O_2, for example, the procedure is as follows. The sample is introduced through the mercury seal at the top of the extraction chamber, with the manometer at ambient pressure. When the valve at the top of the extraction chamber is closed again, the manometer bulb is lowered, drawing a vacuum on the extraction chamber and expanding the gas volume from 25 to 50 times. The extraction chamber is attached to the rest of the apparatus by a flexible, air-tight coupling and to a shaker motor. The sample is vigorously shaken under vacuum for 2 to 3 min, causing nearly all of the dissolved gas to enter the low-pressure gas phase.[1] The shaker is then turned off and the manometer bulb adjusted so that the gas volume contracts to exactly the volume marked (a, Fig. 4.8). The pressure on the manometer is marked as P1. Now a second solution which absorbs oxygen, either sodium sulfite or pyrogallol (1,2,3-trihydroxybenzene), is introduced through the mercury seal, the chamber shaken again, and again the manometer used to return the gas volume to the calibration marks. The pressure difference between the initial and final readings is then used in a simple calculation to find the amount of oxygen originally contained in the sample. A complete

---

[1]The error caused by gas remaining in solution may be calculated by referring to the law of partial pressures and the solubility tables given in Chapter 2. This quantity, generally between 1 and 2%, is incorporated into the correction factors given by Van Slyke and Neill.

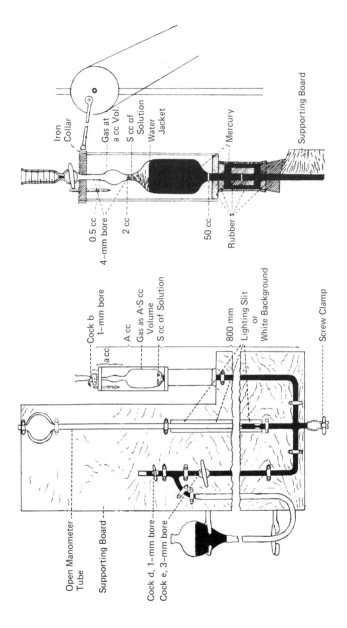

Fig. 4.8 (a) Drawing of the Van Slyke manometric gas analyzer. (b) Detail of the extraction chamber. (Reprinted from Van Slyke and Neill, 1924, J. Biol. Chem., with permission.)

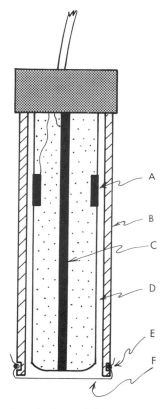

Fig. 4.9 Cross-section of a typical oxygen electrode, showing the silver anode (A), electrode housing (B), platinum cathode (C), electrolyte space (D), and O-ring (E) holding the membrane (F) to the end of the housing.

description of the apparatus, methods for preparing reagents, and points of technique may be found in Van Slyke and Neill (1924).

With some practice and careful technique, the Van Slyke apparatus may yield values accurate to 0.2% with 2.0 ml samples and about 1–2% with 0.2 ml samples. The method is considered rather too tedious by today's standards, and similar accuracy may be obtained with samples of less than 0.1 ml with other methods. For applications in which time is more abundant than equipment money, and where the sample volume is not limiting, the method is still a viable alternative for dissolved oxygen in blood, in particular.

4.4.3. Oxygen Electrodes

Construction of a typical membrane-covered oxygen electrode is shown in Fig. 4.9. It consists of a membrane permeable to oxygen spread over a flat or convex electrode, usually clamped in place with an O-ring. Between the membrane and the surface of the electrode proper there is a layer of electrolyte, which may be stiffened by the addition of agar or methylcellulose. The electrolyte provides a high conductance path between the anode and the cathode, but the anode need not directly face the membrane. The cathodic reaction is summarized by the following equation:

$$O_2 + 2\,H_2O + 4e^- = 4\,OH^- \tag{Eq. 4.4}$$

but involves complex intermediate steps. It is sufficient here to note that an electrical *current* proportional to the oxygen reaching the electrode is generated.

A number of factors influence the magnitude of the current generated by the electrode, including the oxygen permeability of the membrane material, the thickness of the electrolyte layer beneath it, and the total surface area of the cathode (Hale, 1983). When the cathode area is large, the current is easier to amplify, display, and measure, but some undesirable side effects are also present. When the cathode consumption is appreciable, the electrode depletes the oxygen in the neighborhood of the membrane, setting up a diffusion gradient around it. This is manifested as flow sensitivity, and such electrodes require good stirring of the solution over the membrane in order to yield reliable results. If the sample volume is small, the self-consumption of the electrode may significantly lower the reading obtained. In recent years, DC current amplifiers have become much cheaper and easier to build, so the trend is toward very small cathode areas with their correspondingly low consumption, low flow sensitivity, and low current output. One common electrode, the E5046, manufactured by Radiometer-Copenhagen, has a current output of only about 10^{-11} A/torr, but can be used for the measurement of microliter samples with no stirring and no significant electrode consumption error.

For special purposes or just for economy, electrodes may be made easily from the desired metal pairs and some potting epoxy (Mickel et al., 1983). Any of a variety of membrane materials may be used, such as polypropylene or PTFE, but they are highly variable in quality. Microscopic holes and changes in permeability with aging are common problems, so one is probably better off purchasing commercial membrane materials. The metal pairs that may be used are many, and are summarized in Hersch (1964). The electrodes are usually polarized at 0.65 V, which may be accomplished with a mercury battery. (Other battery types are not suitable, as they do not have sufficiently stable age/voltage characteristics.) An alternate electrolyte to be used at a polarizing voltage of 1.05 V has been described (Hahn et al., 1975), and this seems to give the added stability and speed claimed by the authors.

A variety of current-measuring devices may be used to monitor the output of an oxygen electrode. For electrodes with larger current output, a simple hand-held multimeter may suffice. For most practical electrodes, however, the current will be too small for such devices, and an amplifier circuit is necessary. The schematic diagram in Fig. 4.10 will accommodate electrodes with outputs down to as little as 10^{-12} A by adjusting the feedback resistor R according to the formula

$$V_o = -I_f R_f \qquad \text{(Eq. 4.5)}$$

This circuit is also suitable for microelectrodes and is described further in the circuit manufacturer's literature. When such low currents are being measured, however, extreme care must be taken in the layout and cleaning of the circuit in order to prevent leakage pathways from shorting out the feedback resistor. The salt on one's hands will cause current flow on the outside of the resistor, for example, and may provide a much lower resistance pathway than that intended.

Oxygen Microelectrodes. The function of the membrane in the electrodes described above is to control the electrolyte environment over the electrode surfaces, to admit oxygen, and to prevent other reactions from proceeding that eventually "poison" the electrode. In certain environments where the electrolytes are reasonably stable and cross-reactions are not too serious, naked electrodes are feasible. Since the requirements of miniaturization make the membrane and its associated parts for holding it in place undesirable, most microelectrodes are of the naked type. One fairly simple design is shown in Fig. 4.11. The cathode consists of a fine platinum wire (about 0.001″ in diameter) soldered to a copper lead and sealed with a small quantity of epoxy resin into a thin, flexible polyethylene catheter. After the resin has set, the end is cut off cleanly to provide a flat tip with a small cathode area. A separate chlorided silver anode must also be introduced close by. These electrodes may be used intravascularly, and may be employed with the measuring circuit shown in Fig. 4.10 with a feedback resistor of about 100 megohm.

4.4.4. Stripping Oxygen from Solution

Several of the methods for measurement of O_2 in the gas phase may be applied if the dissolved oxygen is first removed, or "stripped," into an oxygen-free carrier gas. The simplest way to accomplish this is to bubble nitrogen or any other suitable gas through the solution containing oxygen. The resulting partial pressure gradient between the solution and the gas phase will cause the oxygen to move down its diffusion gradient from the solution

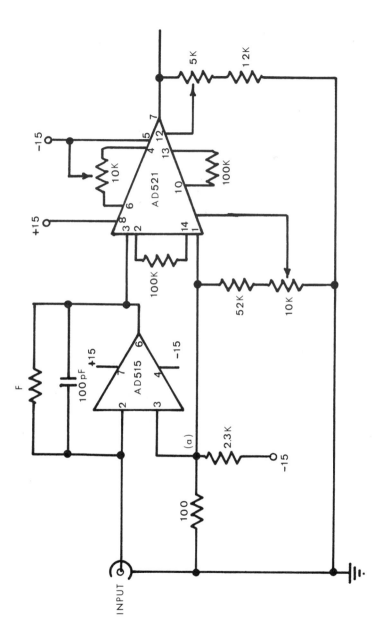

Fig. 4.10 Complete schematic diagram of an oxygen electrode circuit, including a high input impedance current-to-voltage converter stage (AD515, Analog Devices, Inc.) and an additional level-shifting and gain stage (AD521, Analog Devices, Inc.). The electrode's zero current is balanced to zero with the 10K variable resistor, the amplifier zero voltage nulled with the resistor connected to pins 4 and 6 of the AD521, and the overall gain controlled with the 5K resistor connected to pins 7 and 12. The polarizing voltage (measured at a) is set to -0.65 V, but it can easily be set to other values by changing the resistor values. Various electrode currents can be accommodated by adjusting the value of the feedback resistor (F).

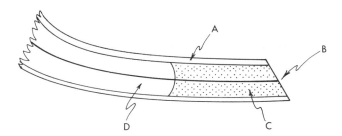

Fig. 4.11 Cross-section of an oxygen microelectrode. A fine platinum wire (B) is potted in a small polyethylene catheter tube (A) with epoxy or other suitable material (C). The remainder of the electrode bore is back-filled with KCl (D).

into the gas. As the gas is continually removed, the process will continue until the oxygen is completely removed from the solution. The rate at which oxygen is stripped in this manner is a function of the original content and the surface area for exchange, which may be increased greatly by a very small bubble size, and to a lesser extent by increasing the gas flow rate. Unfortunately, in many solutions of physiological interest bubbling causes foaming, especially when proteins and lipids are present. In those cases, the stripping gas may be passed over the surface of a rapidly stirred thin layer of solution. Extraction is slower than with bubbling, but still practical.

The carrier gas stream, which now contains the oxygen removed from solution, may be led into various sorts of analyzers, including galvanic cells, paramagnetic analyzers, or gas chromatographs. Before introduction, the water vapor may have to be removed with a drying trap, and in the case of the gas chromatographs, the stream may have be sampled discretely, since the method precludes continuous introduction of sample material. The time course of oxygen removal from solution is a fairly complex function, and it is usually necessary to integrate the signal to obtain an area measurement.

LITERATURE CITED

Hahn, C. E. W., A. H. Davis, & W. J. Albery. 1975. Electro-chemical improvement of the performance of P_{O_2} electrodes. Respir. Physiol. 25: 109–133.

Hale, J. M. 1983. Factors influencing the stability of polarographic oxygen sensors. *In* Polarographic Oxygen Sensors: Aquatic and Physiological Applications. E. Gnaiger & H. Forstner, eds. Springer-Verlag, New York. pp. 3-17. (This volume has articles on many aspects of oxygen electrode design and application.)

Hersch, P. 1964. Galvanic analysis. Adv. Anal. Chem. & Instr. 3: 183–249.

Mickel, T. J., L. B. Quetin, & J. J. Childress. 1983. Construction of a polarographic oxygen sensor in the laboratory. *In* Polarographic Oxygen Sensors: Aquatic and Physiological Applications. E. Gnaiger & H. Forstner, eds. Springer-Verlag, New York. pp. 81–85.

Scholander, P. F. 1942. A micro-gas-analyzer. Rev. Sci. Instr. 13: 264–266.

Scholander, P. F. 1947. Analyzer for accurate estimation of respiratory gases in one-half cubic centimeter samples. J. Biol. Chem. 167: 235–250.

Van Slyke, D. D., & J. M. Neill. 1924. The determination of gases in blood and other solutions by vacuum extraction and manometric measurement. I. J. Biol. Chem. 61: 523–573.

CHAPTER 5

CARBON DIOXIDE MEASUREMENT

5.1. BASIC PROPERTIES

Atomic weight:	44.01
Melting point:	$-41.6°C$
Triple point:	$31.1°C$, 838 psig
Density, $0°C$:	1.832 g L^{-1}

Miscellaneous. Colorless, odorless, stable gas which is chemically inert under normal conditions. Absorbs in the infrared. Sublimes directly from solid to gas at $-41.6°C$. At the triple point the solid, liquid, and gas forms may all co-exist. Only gas is present at any temperature above the critical point.

5.2. THE CHEMISTRY OF CO_2 IN WATER

Gaseous CO_2 is quite soluble in water (see Chapter 2), but its behavior in aqueous solution is considerably complicated by its ability to react chemically with water. There are a series of reactions, the first of which forms *carbonic acid*:

$$CO_2 + H_2O \longrightarrow H_2CO_3 \qquad \text{(Eq. 5.1)}$$

An expression for the equilibrium constant for this reaction may be written as follows:

$$L = [CO_2]/[H_2CO_3] \qquad \text{(Eq. 5.2)}$$

The equilibrium constant for this reaction is very large, with less than 0.5% of the CO_2 being transformed into carbonic acid. The reaction of CO_2 with water proceeds to a much greater extent than predicted by (5.2), however, due the stepwise dissociation reactions of carbonic acid:

$$H_2CO_3 \longrightarrow H^+ + HCO_3^- \qquad \text{(Eq. 5.3)}$$

and

$$HCO_3^- \longrightarrow H^+ + CO_3^= \qquad \text{(Eq. 5.4)}$$

The dissociation constants for these two reactions are

$$K_1 = [H^+][HCO_3^-]/[H_2CO_3] \qquad \text{(Eq. 5.5)}$$

$$K_2 = [H^+][CO_3^=]/[HCO_3^-] \qquad \text{(Eq. 5.6)}$$

In many situations it is convenient to describe the total concentration of all forms of (dissolved plus chemically combined) CO_2, and this total concentration, C_{Tot} is

$$C_{Tot} = \alpha P_{CO_2} + [H_2CO_3] + [HCO_3^-] + [CO_3^=] \quad \text{(Eq. 5.7)}$$

By combining several of the above equations and taking activities rather than concentrations into account, the relationship between pH and various forms of CO_2 may be conveniently expressed by the Henderson–Hasselbalch equation:

$$pH = pK_1' + \log((C_{Tot}/\alpha P_{CO_2}) - 1) \qquad \text{(Eq. 5.8)}$$

The right-hand term may be rearranged using (5.7): $[H_2CO_3]$ is always very small, and at physiological pH when the $[CO_3^=]$ is insignificantly low, an alternative form is

$$pH = pK_1' + \log((HCO_3^-)/\alpha P_{CO_2}) \qquad \text{(Eq. 5.9)}$$

The value of pK_1' depends on the pH and the ionic strength, so it is a "pseudo-constant" employed for convenience, but in physiological situations it is approximately 6.1. The value for pK_2' (representing the carbonate dissociation step) is around 9, and so the formation of carbonate need only be taken into account at high pH values and low temperatures. A more complete discussion of the chemistry of the carbonic acid system may be found in many texts, but especially good treatments are those by Hills (1973) and Albers (1970).

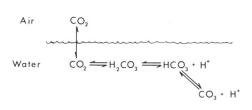

Fig. 5.1 Equilibrium diagram for CO$_2$, showing diffusive equilibrium of CO$_2$ between air and a liquid phase, as well as subsequent reactions of CO$_2$ with water and dissociation of the resultant carbonic acid.

From the standpoint of conducting analyses of CO$_2$ in aqueous solution, several features of the chemistry are useful to remember. Acidification favors conversion of all the forms to dissolved CO$_2$ gas (Fig. 5.1), and at pH values below 4, less than 1% of the total exists in any other form. Thus analytical methods that are based on detecting the CO$_2$ as a gas usually involve acidification. At high pH values the formation of carbonate is heavily favored, such that at pH 12 and above, less than 1% exists in any form other than carbonate. There are a number of analytical methods that depend on carbonate formation, particularly the formation of insoluble carbonates. In the remainder of this chapter, various methods are discussed for measuring the total CO$_2$ content, C_{Tot}, and P_{CO_2}; methods for measuring pH are discussed elsewhere (Chapter 6).

5.3. USEFUL REACTIONS OF CO$_2$

Besides the carbonic acid system reactions discussed in the previous section, there are a number of reactions of CO$_2$ that are useful analytically. CO$_2$ reacts with most bases to form carbonates, and with the hydroxides of alkali metals the carbonates are generally insoluble or very slightly soluble in water. At high pH, for example, the reaction

$$Ca(OH)_2 + CO_2 = CaCO_3 + H_2O \qquad \text{(Eq. 5.10)}$$

figures importantly in calcification and is relevant to such diverse processes as hardening of skeletal structures and formation of coral reefs. The same reaction with Ba instead of Ca produces a precipitate that is even less soluble and is useful as an analytical tool. The oxides of alkali metals may also react directly with CO$_2$, possibly via the mediation of a small adsorbed water layer on the crystal surface. The reaction with Li$_2$O is strongly exergonic (heat-yielding), and the quantitative measurement of heat evolution has been used as an analytical tool for gaseous CO$_2$.

5.4. MEASUREMENT OF CO_2 IN GASES

5.4.1. Manometric Methods

A variety of manometric methods for CO_2 measurement have been described over the past century or so, mostly based upon measurement of the reduction of gas volume after absorption of the CO_2 into an alkaline solution of some sort. For example, the Scholander apparatus described in Section 4.3.1. is still in use. With practice and careful technique, manometric methods can be quite accurate and reliable, but they are subject to all the error sources of manometric methods generally. The most important of these are volumetric measurement error, temperature error, incomplete reaction (absorption), and leakage of the apparatus. They have the advantage of low cost, but are time-consuming.

5.4.2. Gravimetric Methods

The reaction of CO_2 with alkali metal oxides may be utilized in a very straight-forward method for CO_2 analysis. A common application is in the measurement of respiratory CO_2 production by animals. The expired air stream is led first through a drying tube or similar water vapor trap, then through a container filled with CaO. The CaO is usually granular, with a very high surface area so that the absorption and reaction are virtually 100% complete. By simply weighing the tube at the beginning and end of the measurement interval, the quantity of CO_2 collected is directly measured. Various commercial CO_2 absorbents are available, often with an indicator added to show when the material is exhausted.

An equally suitable method is to bubble the gas containing CO_2 through a solution of an alkali metal hydroxide, barium being the usual choice due to the extremely low solubility of its carbonate. The carbonate formed is later collected by filtration or other means, dried, and weighed. Quantitative recoveries are possible, but contamination with atmospheric CO_2 must be carefully guarded against.

In a slight variant of this method, the alkali metal hydroxide remaining after CO_2 absorption may be measured by titration. Comparison of this titration value with titration of a blank aliquot allows calculation of the CO_2 absorbed by difference. This method obviates collection of the precipitate, and is somewhat easier to perform accurately.

5.4.3. Infrared Analysis

The carbon–oxygen bond of CO_2 is susceptible to excitation by infrared radiation and absorbs in the 4.3 and 15.6 μm wavelength regions. This property is the basis for infrared CO_2 analyzers. The absorption of the infrared radiation is relatively weak, so a long path length of the beam through the gas must be provided, on the order of 1 m. Rather than actually constructing a cell of this length, the beam is usually reflected several times through the length of a shorter cell with mirrors before entering the detector. The gas volume required is usually fairly large, but the method is well suited to continuous measurements and has a high sensitivity. None of the other atmospheric gases have any infrared absorption, so there are no significant interferences to be concerned with.

5.4.4. Mass Spectrometry

Mass spectrometry was described in detail in Chapter 4, and that discussion applies to CO_2 as well as to O_2. The method may be applied either to single samples or on a continuous stream basis and has the advantage of a small sample requirement for single samples. In the continuous stream mode, the sample requirement depends on the speed of response desired, as with the analysis of O_2. For rapid responses, say in breathing cycle analyses, a fairly high inlet flow may be required. Small variations of inlet flow are usually corrected for by measuring the ratio of CO_2 to nitrogen, since the latter is practically constant in physiological studies. Mass spectrometry is particularly useful when employing stable isotopes of carbon in metabolic or respiratory studies, since the mass spectrometer easily distinguishes between $^{12}CO_2$ and $^{13}CO_2$, masses 44 and 45. As with O_2 analysis, the major disadvantages are the size, complexity, and cost of the apparatus.

5.4.5. Gas Chromatography

The analysis of CO_2 in gas mixtures, particularly in atmospheric gases, is easily carried out by gas chromatography. The general method is described in Chapter 4, and several "molecular sieve" types of column materials, as well as Poropak Q, are suitable stationary phases for the separation. The thermal conductivity of CO_2 is similar to that of oxygen (Table 4.1), so thermal conductivity detection with helium as a carrier gas is also the best combination for gas chromatography analysis of CO_2.

5.4.6. Conductometric Methods

As pointed out in Section 5.2, high pH favors the formation of $CO_3^=$ ion in solution. If CO_2 is absorbed into a weak NaOH solution, the ion pair $NaCO_3^-$ forms to a considerable extent, and the electrical conductivity of this ion pair is significantly less than that of Na^+ and OH^- ions. Thus, if the electrical conductivity of the NaOH solution is measured before and after addition of CO_2, a quantitative analysis may be performed by comparing the conductivity change from an unknown sample with that of a known standard. Methods described in the literature are based upon a variety of physical arrangements of the gas sample and the absorbing alkali, and may be adapted either for discrete sample analysis or for continuous stream measurements (Maffly, 1968; Bruins, 1963). Exremely high sensitivity may be achieved, extending the measurement capability to the nanoliter (10^{-9} L) range. The electrical conductivity of solutions is quite temperature sensitive, so careful control is necessary. No commercial instrument based on this method is available specifically for gas measurement, so the cost of the apparatus must include custom glassware construction.

5.5. PARTIAL PRESSURE MEASUREMENT IN LIQUIDS

5.5.1. Electrodes

The CO_2 electrode was invented by Stow *et al.* (1957) and is shown in schematic form in Fig. 5.2. It consists of a pH electrode enclosed in a housing filled with a buffer electrolyte and separated from the sample by a membrane permeable to CO_2. The buffer is usually a mixture of NaCl and $NaHCO_3$. From (5.9) it may be seen that the relationship between the partial pressure of CO_2 (P_{CO_2}) and pH is linear on a log scale, at least to the extent that [HCO_3^-] does not change (Fig. 5.3). When the electrolyte inside the electrode body is equilibrated with the external sample, its pH is then a direct log-linear function of the sample P_{CO_2}. The impetus for development of the electrode was the need for measurements in human clinical physiology, i.e., at relatively high P_{CO_2} values and a temperature of 37°C. At lower temperatures and P_{CO_2} values, however, several limitations become troublesome. The diffusion of CO_2 through the membrane and through the volume of electrolyte contained between the membrane and the electrode face is slow if the total distance is more than 100 μm or so or if the membrane permeability is low. Thinner and more permeable membranes may partially offset this problem, and 4 μm thick silicone rubber is a preferred membrane type. For higher temperatures, Teflon is more stable but much slower. The thickness of the electrolyte layer over the face of the electrode is usually fixed by either a piece of nylon gauze or tissue paper, and this can also be made thinner, but only within limits. Finally, the electrode response time is limited by the rate of the hydration–dehydration reaction of CO_2 to HCO_3^- and H^+ ions, which has a half-time of approximately 8 sec at 37°C, but 330 sec at 0°C. The combined effects of membrane resistance, electrolyte volume, and reaction time yield an electrode response time on the order of 30 to 45

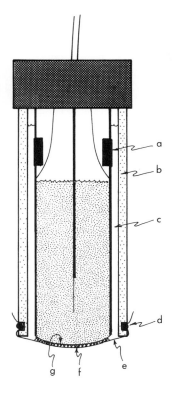

Fig. 5.2 Cross-section of a CO_2 electrode, showing the Ag/AgCl reference (a), the electrode housing (b), the electrolyte space (c), the O-ring (d) holding the membrane (e) in place, a thin layer of gauze between the membrane and the electrode face (f), and the thin pH-sensitive glass tip of the electrode (g).

sec at 37°C but 12 to 15 min at 10°C, which is unacceptable for many applications. At low values of P_{CO_2} the response time is also lengthened, and electrode drift becomes a problem.

The CO_2 electrode may be used with any good pH meter and an empirical calibration curve like Fig. 5.3 constructed using known-concentration calibration gases. Alternatively, various commercial special-purpose meters may be used which automatically do the log-linear conversion and provide a direct display of partial pressure.

5.5.2. Mass Spectrometry

Although the mass spectrometer is usually applied to the analysis of gases, it can be used for direct analysis of dissolved gases in aqueous solution. Instead of employing a mechanical "leak" inlet, a semi-permeable membrane may be used directly as the leak. That is, a small metal tube, evacuated inside

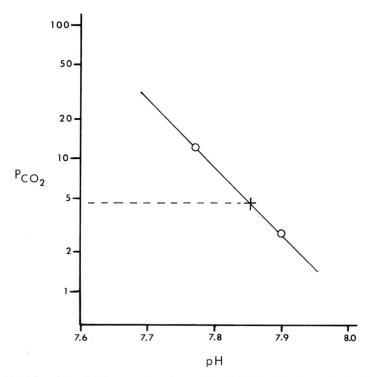

Fig. 5.3   The relationship between P_{CO_2} (log scale) and pH in a bicarbonate-buffered solution. If the pH of an unknown solution is measured (point x) and the solution is subsequently equilibrated with two known gases (yielding the open points), the relationship can be used to estimate the P_{CO_2} of the unknown solution, as shown by the dotted line (Astrup method).

and fitted with a membrane at its tip, may be introduced directly into the liquid sample. The CO_2 will pass through the membrane at a rate proportional to its partial pressure, provided there is no depletion of CO_2 in the region surrounding the membrane. A commercial instrument with this capability was offered a few years ago (by Medspect), but various technical problems led to its discontinuance. Mass spectrometry offers the potential advantages of continous sampling of other gases as well as CO_2, and relative insensitivity to temperature. Perhaps with further development of the leak technology, including membranes less sensitive to protein contamination and more flexible materials, mass spectrometry will become more attractive than it is at present.

5.5.3. Indirect (Calculation) Methods

Probably the most common methods for measurement of P_{CO_2} are indirect calculation methods involving the application of the Henderson–Hasselbalch equation (Eq. 5.9). That is, if one measures the total CO_2 and the pH, and if the pK_1' for the system is known, the P_{CO_2} can be calculated. Various authors give values for pK_1' as a function of temperature, ionic strength, and pH in the form of tables or nomograms (Nicol *et al.*, 1983; Sigaard-Andersen, 1974; Maas *et al.*, 1971). The error limits of such calculations are not always easy to assess, since they are logarithmic and may be multiplicative.

A second indirect method is the so-called Astrup method. The method takes advantage of the relationship between P_{CO_2} and pH shown in Fig. 5.3 and entails several steps. First, the pH of the unknown solution is measured. Then aliquots of the sample are equilibrated with two gases of known P_{CO_2}, and the pH of each equilibrated aliquot is measured. Using the calibration points to draw a line, as in Fig. 5.3, the P_{CO_2} of the original sample may be scaled from the plot. The measurement of pH may be made with reasonable precision (see Chapter 6), but various errors in the equilibration step are possible, including incomplete reaction, evaporative concentration, etc.

5.6. MEASUREMENT IN LIQUIDS: TOTAL CONTENT

5.6.1. Stripping and Extraction

The general considerations for removing dissolved gas from liquids have been discussed in Section 4.4.4, but with CO_2 there are the additional problems of a much greater solubility in the liquid and the necessity to convert the various chemical species to the dissolved gas form. The conversion can be accomplished by acidification, but it is usually desirable to employ one of the non-volatile acids, such as lactic, phosphoric, or sulfuric acid. If sulfuric acid is used, it must not be at concentrations sufficient to oxidize organic materials in the solution, thereby generating CO_2 from that source. The extraction may be performed by evacuation, as in the Van Slyke apparatus (Section 4.4.2), or more commonly using a CO_2-free carrier gas. Virtually any convenient gas will suffice, but it is always good practice to pass it through a CO_2 trap first. Good materials for CO_2 removal are Ascarite or a concentrated KOH solution. As with O_2 extraction, either fine bubbling through the solution or rapid stirring of a thin layer may be employed, depending on the tendency of the liquid to foam when bubbled.

5.6.2. Manometric Methods

The Van Slyke apparatus described in Section 4.4.2 is the best known of the manometric methods for total CO_2 content in aqueous samples, and the general procedure is almost identical to that described for analysis of oxygen. The only difference is that a strong alkali, usually NaOH, is used to absorb the CO_2, after which the volume change is measured. Both O_2 and CO_2 are

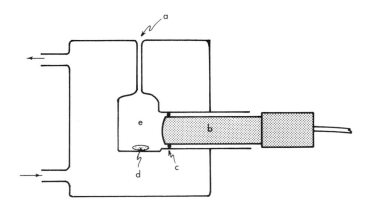

Fig. 5.4   The electrode and cuvette method for measurement of total dissolved + combined CO_2 (C_{Tot}) in aqueous solutions. (Modified by permission from Cameron, 1971.)

routinely measured by this method. For application to blood and serum, some special anaerobic techniques are required (Chapter 10).

5.6.3. The Electrode/Cuvette Method

As pointed out above (Section 5.2), acidification drives all the CO_2 equilibrium reactions toward dissolved CO_2 gas (Fig. 5.1), so a combination of acidification and P_{CO_2} measurement using a CO_2 electrode may be used for the measurement of C_{Tot}. This method, originally described by me (Cameron, 1971), employs the apparatus shown in Fig. 5.4. The CO_2 electrode is enclosed in a temperature-jacketed chamber of about 1 to 2.5 ml volume, and maintained at a temperature between 30 and 40°C for a fast response. The chamber is filled with dilute acid at the same temperature and the P_{CO_2} is measured, often with a bit of $NaHCO_3$ or K_2CO_3 added to bring the meter on scale. When the electrode reading is stable, either a known-concentration standard or an unknown is introduced into the acid chamber through the small bore tube with a microliter syringe. The acid causes conversion of the various CO_2 forms to dissolved gas, and the electrode registers the increase in P_{CO_2}. Comparison of the increase resulting from known standards and unknowns allows calculation of the C_{Tot} of the unknowns. The method is applicable to samples of about 20 to 100 μl volume with concentrations ranging from about 5 to 60 mM L^{-1}. Samples with lower concentrations, such as natural waters, may be analyzed by using a large sample volume to fill the chamber and a small added volume of relatively concentrated acid.

While this method has found fairly widespread use, it has certain drawbacks, particularly those associated with CO_2 electrodes. Even at elevated temperatures, there is often a considerable drift, which necessitates frequent calibration. Microscopic holes in membranes may allow diffusion of acid into the electrode, requiring a lengthy process of changing membranes and getting the electrode's temperature stabilized again. Also, the log-linear nature of the P_{CO_2}/pH relationship means that it is desirable to calibrate the P_{CO_2} electrode with a known gas, which requires the apparatus for supplying this gas. The combined cost of the electrode, gas supply, meter, chamber and stirrer, etc., may be considerable.

5.6.4. Conductometric Methods

The general principle of conductometric measurement of CO_2 has been described in Section 5.4.6, and this principle combined with acid extraction is the basis of a commercial instrument for C_{Tot} measurement, shown in Fig. 5.5 (Cameron, 1982). Samples to be analyzed are injected into either of two extraction chambers, one a bubbling type for non-foaming liquids, the other a stirred type for liquids that do foam when bubbled. A CO_2-free nitrogen stream carries the liberated CO_2 into a glass column, which acts as both a CO_2 absorber and a bubble lift pump. As the bubbles rise, they cause the liquid to circulate from the surrounding bath, first through one conductivity detector cell, then through the spiral absorber section, and finally through a second cell. The liquid in the bath is dilute NaOH, which absorbs CO_2 with an attendant change in electrical conductivity. The absorbed CO_2 produces a conductivity peak similar to a peak from a gas chromatograph, and the total area of the peak is directly proportional to the total CO_2 of the sample. The instrument is calibrated by injection of known-concentration standards.

This method is applicable to a variety of liquids, ranging in sample volume from 5 to 200 μl, and in concentration from 1 to 60 mM L^{-1}. Temperature changes are minimized by the use of the differential detection method; i.e., changes in the conductivity due to bath temperature are equally sensed by both cells and cancel out. Analyses may be carried out in 1.5 to 2.5 min. The instrument may also be used for gas samples, provided that the larger volumes required (0.2 to 1 ml) are injected over a several-second interval to avoid large changes in the carrier gas flow rate.

5.6.5. Gas Chromatography

One requirement of good chromatographic analysis is that the sample to be analyzed be introduced onto the column in as small a "package" as possible. In order to apply this method to analysis of the total CO_2 content in liquid samples, it is necessary to collect the extracted CO_2, which comes out of solution in a very gradual fashion, and to inject it as a bolus onto the gas chromatography column. One scheme for doing this is shown in Fig. 5.6.

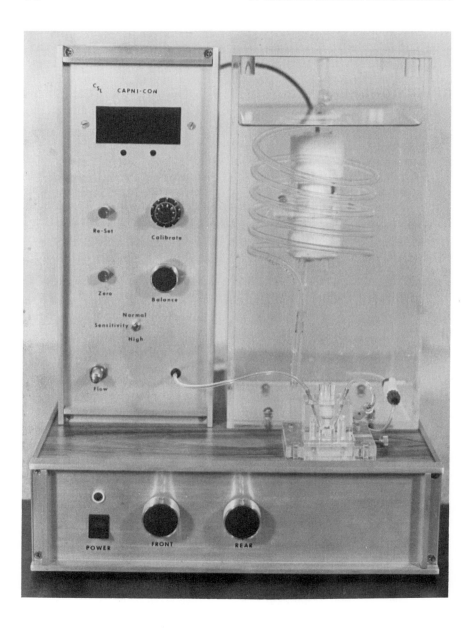

Fig. 5.5 The Capni-Con, an instrument for the conductometric measurement of total (dissolved + combined) CO_2 in aqueous samples. (Courtesy Cameron Instrument Co.)

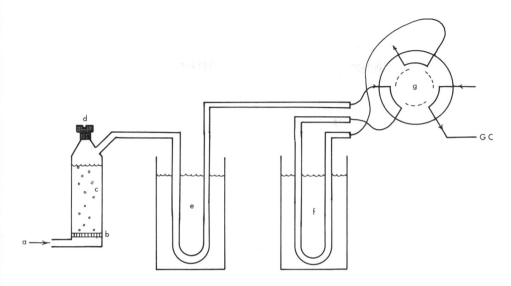

Fig. 5.6 Schematic diagram of a method for measurement of total CO_2 (C_{Tot}) in aqueous solutions using a gas chromatograph (see text for details).

It involves collection of water vapor in a dry ice/methanol trap and collection of the CO_2 evolved from the sample in a liquid N_2 trap. This second trap is actually a sample loop connected to a sampling valve on the gas chromatograph (usually a standard accessory). After the collection is complete, the liquid N_2 is removed, allowing the CO_2 to change back from solid to gas form, and the valve is switched to introduce the gas into the gas chromatograph. In theory this is a very good method, but in practice it may be messy and rather expensive, requiring a ready supply of dry ice and liquid nitrogen.

5.6.6. Calorimetry

The exothermic nature of the reaction between CO_2 and Li_2O was mentioned above (Section 5.3) and forms the basis of a commercial instrument for the measurement of very small quantities of total CO_2 in liquids (Vurek *et al.*, 1975; Picapno-Therm). The instrument employs a small glass extraction chamber, with phosphoric acid used to convert the combined forms to dissolved CO_2. The total heat generated in a small crystal of Li_2O is measured with a bridge circuit and an integrator. The principal advantage of the method and apparatus is the ability to measure picomole quantities, which has proved important in some areas, notably in studies of isolated kidney tubules.

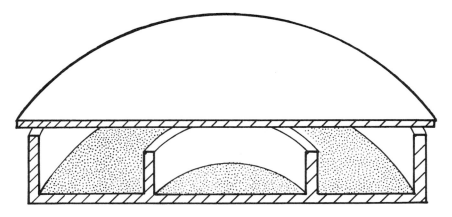

Fig. 5.7 Cross-section of a Conway microdiffusion vessel. (Conway, 1957.)

5.6.7. Microdiffusion Methods

The general idea of microdiffusion analysis is to convert the species of interest to a gas in one section of a closed vessel and to trap the diffusing gas in another section by chemical means. For CO_2 analysis, the microdiffusion method as described by Conway (1957) is based on the vessel shown in Fig. 5.7. The center well of the vessel contains $Ba(OH)_2$ and the outer well a strong acid. The chamber must be sealed under CO_2-free nitrogen and the sample added without excessive exposure to air. When the sample to be analyzed is introduced into the outer well and swirled gently to mix it with the acid, the CO_2 is converted to the dissolved gas form, as with the other methods discussed. The dissolved CO_2 then tends to equilibrate with the gas phase above the acid, which brings it into contact with the $Ba(OH)_2$ in the center well. After some suitable equilibration time, all of the CO_2 becomes trapped by reaction in the base to insoluble $BaCO_3$. The vessel is then opened under a nitrogen stream and the remaining $Ba(OH)_2$ measured by titration. This method, although somewhat tedious, is quite inexpensive and will yield results of high accuracy, provided care is taken in sealing the vessels and in not exposing reagents or vessels to the air.

5.7. CONCLUSIONS

The foregoing discussion should serve to illustrate the great variety of methods that may be employed for the measurement of CO_2 in either gases or liquids. The various methods are based on several of the physical and chemical properties of CO_2, and the variety available presents something of a problem for anyone trying to choose the most appropriate method for a given application. Some of the factors which need to be taken into account are the total cost of equipment and supplies, the number of analyses to be performed, the required accuracy, the time available for analyses, and the general level of skill of the operator. Many of the methods may be modified in various ways to suit a particular application, but there is probably a suitable method for almost any conceivable situation.

LITERATURE CITED

Albers, C. 1970. Acid–base balance. *In* Fish Physiology. W. S. Hoar & D. J. Randall, eds. Academic Press, New York. Vol. IV, pp. 173–208.

Bruins, E. H. 1963. Determination of carbon dioxide using a high frequency oscillator. Anal. Chem. 35: 934–936.

Cameron, J. N. 1971. A rapid method for determination of total carbon dioxide in small blood samples. J. Appl. Physiol. 31: 632–634.

Cameron, J. N. 1982. Carbon dioxide measurement system. U.S. Patent #4,321,545.

Conway, E. J. 1957. Microdiffusion Analysis and Volumetric Error. Crosby Lockwood, London.

Hills, A. G. 1973. Acid–Base Balance: Chemistry, Physiology, Pathophysiology. Williams & Wilkins, Baltimore. 381 pp.

Maas, A. H. J., A. N. P. van Heijst, & B. F. Visser. 1971. The determination of the true equilibrium constant (pK_1) and the practical equilibrium coefficient for the first ionization of carbonic acid in solutions of sodium bicarbonate, cerebrospinal fluid, plasma and serum at 25 and 38°C. Clin. Chim. Acta 33: 325–343.

Maffly, R. H. 1968. A conductometric method for measuring micro-molar quantities of carbon dioxide. Anal. Biochem. 23: 252–262.

Nicol, S. C., M. L. Glass, & N. Heisler. 1983. Comparison of directly determined and calculated plasma bicarbonate concentration in the turtle *Chrysemys picta bellii* at different temperatures. J. Exp. Biol. 107: 521–525.

Sigaard-Andersen, O. 1974. The Acid–Base Status of the Blood. (4th ed.). Munksgaard, Copenhagen.

Stow, R. W., R. F. Baer, & B. F. Randall. 1957. Rapid measurement of the tension of carbon dioxide in blood. Arch. Phys. Med. Rehabil. 38: 646–650.

Van Slyke, D. D., & J. M. Neill. 1924. The determination of gases in blood and other solutions by vacuum extraction and manometric measurement. I. Complete description of apparatus and techniques. J. Biol. Chem. 61: 523–573.

Vurek, G. G., D. G. Warnock, & R. Corsey. 1975. Measurement of picomole amounts of carbon dioxide by calorimetry. Anal. Chem. 47: 765–767.

CHAPTER

6

pH

6.1. THE CONCEPT OF pH

Water is the universal medium for life, so an understanding of its chemistry is essential for any biological study. The water molecule, H_2O, dissociates to a small extent according to the reaction

$$H_2O \longleftrightarrow H^+ + OH^- \qquad \text{(Eq. 6.1)}$$

and a dissociation constant for this reaction may be written as follows:

$$Kw = [H^+][OH^-] \qquad \text{(Eq. 6.2)}$$

The dissociation constant K_w is about equal to 10^{-14}, depending on the temperature and ionic strength of the solution, and since $[H^+]$ and $[OH^-]$ must be equal in pure water, the concentration of H^+ is half of K_w, or about 10^{-7} mol L^{-1}. (H^+ ions are normally present as the hydrated form H_3O^+, but for the sake of simplicity, the notation H^+ will be used.) Addition of H^+ from strong acids will change the concentration of OH^- in a reciprocal fashion, so that their product remains equal to K_w. For example, in a 0.1 N solution of HCl the $[H^+]$ would be 10^{-1}, and the $[OH^-]$ would then be 10^{-13} mol L^{-1}.

The ideal solution laws to which the above discussion conforms are not strictly adhered to by real solutions. In non-ideal solutions, it is the *activity* of the various species in solution that is important, so rather than $[H^+]$, the operational parameter is the activity of the H^+ ion, or a_{H^+} (see Section

2.8). It is also more convenient to use logarithmic units, and the pH is defined as:

$$pH = -\log a_{H^+} \qquad\qquad (Eq.\ 6.3)$$

The use of this convenient manner of describing H^+ ion activity requires a good grasp of logarithmic units. As the H^+ activity rises, the pH falls, for example, and an increase in pH by 0.3 units means halving of H^+ activity and doubling of OH^- activity. A change in one pH unit means a change in $[a_{H^+}]$ by a factor of 10.

An absolute standard for pH has been rather difficult to agree upon, so now the standard is set as the electrical potential measured from a precisely described electrode chain under a defined set of solution conditions. The fact that H^+ activity and not concentration is measured, the somewhat arbitrary calibration of the scale, and some consideration of experimental error distribution has significance for a rather long-standing argument. The argument is whether, in the literature, it is more appropriate to express results using pH units or $[H^+]$ units, and particularly whether statistical error limits (standard deviation and standard error) should be calculated using one or the other. Calculating the standard error of pH measurements by first converting to the H^+ concentration will yield asymmetric numbers when converted back to pH or activity units, due to the logarithmic nature of the pH scale. It should be clear from the definition and activity aspects that the use of pH is more appropriate than $[H^+]$, but a decision might also rest upon whether the error distribution is normal on a log or a linear basis. Seldom are a given set of data complete enough to test for the difference, but the weight of the evidence indicates a log-normal error distribution, or at least does not contradict such an assumption (Boutilier & Shelton, 1980), so this criterion too dictates the use of pH. Presentation of results as H^+ concentration (not activity) is necessarily fiction, since this quantity is never measured.

The effect of temperature on the dissociation of water has been frequently overlooked but is quite important in physiology. Table 6.1 gives values for K_w over the range of 0 to 50°C. It is convenient here to introduce another concept, that of pK, which is equal to the negative logarithm of K. The pK value represents the point at which the dissociation is half complete, i.e., $[H^+]$ exactly equals $[OH^-]$. For water this is the *neutral* pH, which from Table 6.1 may be seen to be 7.00 at 24°C. It may also be seen from the table that the pH of neutrality varies between 7.47 and 6.63 over the physiological temperature range, so the idea that pH 7 is neutral must be modified to include temperature as a factor.

TABLE 6.1
The Dissociation Constant for Water (K_w) at Various Temperatures, Along with the Value of pN $(= -\log \frac{1}{2}K_w)^a$

Temp. (°C)	K_w	pN
0	14.9435	7.4718
5	14.7338	1.3669
10	14.5346	7.2673
15	14.3463	7.1732
20	14.1669	7.0835
24	14.0000	7.0000
25	13.9965	6.9983
30	13.8330	6.9165
35	13.6801	6.8401
40	13.5348	6.7674

^{a}Data from the *Handbook of Chemistry and Physics*.

6.2. BUFFERING

The idea and the effects of buffering may be seen by using the examples in Fig. 6.1. On the left is a titration curve of a strong acid, HCl, titrated with a strong base, NaOH. A strong acid is one whose dissociation constant is very large; in the case of HCl:

$$K = [H^+][Cl^-]/[HCl] \qquad \text{(Eq. 6.4)}$$

A large K value means that the acid is almost completely dissociated into H^+ and Cl^- ions and that the concentration of undissociated HCl is very small. As the H^+ ions are titrated by the OH^- from NaOH, the pH of the solution changes relatively little until the concentrations near equality. To use some numerical examples, if we start with 1.1 N HCl in a 1 L solution and add NaOH by 0.2 Equiv increments, after five additions there will still be 0.1 Equiv L^{-1} of H^+ ions, and the pH will be -log(10^{-1}), or 1.0. After the next increment of NaOH, there will be an excess of OH^- ions of 0.1 Equiv L^{-1}, so that $[OH^-] = 10^{-1}$ mol L^{-1}, $[H^+] = 10^{-13}$ mol L^{-1}, and pH = 13. This provides the "square" titration curve shown in Fig. 6.1a and shows no buffering.

A weak acid, on the other hand, has a small K value, and according to (6.4), will be only partly dissociated. The Henderson–Hasselbalch equation

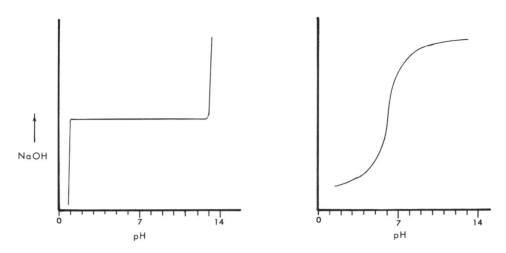

Fig. 6.1 (Left) A titration curve for a strong acid showing the sharp transition from acid to alkaline pH. (Right) A titration curve for a weak acid showing the more gradual transition due to buffering.

is useful for describing the relationship between pH and acid dissociation. For acetic acid (HAc)

$$HAc \longleftrightarrow H^+ + Ac^- \qquad\qquad (Eq.\ 6.5)$$

and

$$pH = pK + \log([Ac^-]/[HAc]) \qquad (Eq.\ 6.6)$$

As with water, the pK gives the point at which the acid is exactly half dissociated, since the right-most term is equal to 1 when the acid is half dissociated, and the log of 1 is zero. Now when the acid is titrated with a strong base, the dissociation reaction supplies more H^+ ions gradually as OH^- is added, producing a much more gradual pH transition like that shown in Fig. 6.1b. This tendency to resist change in pH with addition of strong acid or base is called *buffering*, and depends upon the concentration of the buffer substance and the relationship of the pH of the solution to the buffer's pK. From Fig. 6.1b it may be seen that the slope is greater, and hence the buffering is greater, near the pK value of the acid (4.74 for HAc). For physiological studies, the significance of this is that only buffers with pK values within about 1.5 units of the physiological range of pH will have important buffer action.

A wide range of buffers is in common use, and these may be either purchased from various suppliers or made up from their constituents. Some

Fig. 6.2 Typical pH measurement equipment, with a glass electrode (right) which incorporates the calomel reference, a beaker of solution, and a modern pH meter.

common buffer recipes are given in Appendix 5, including data on the effects of temperature, and tables for some National Bureau of Standards buffers.

6.3. MEASURING pH IN BULK SOLUTIONS

6.3.1. pH Electrodes

In 1906 a soft soda-lime glass was discovered by Cremer that was H^+ permeable, and the first glass pH electrodes were constructed. The general scheme has changed little and is shown in Fig. 6.2., although many refinements in electrode construction and meter design have been made. When the glass electrode is immersed in a solution whose pH is to be measured, the H^+ ions at first diffuse across the pH-sensitive glass membrane until an opposing electrical potential is developed. The potential developed and the external H^+ activity are related by the Nernst equation:

$$E = E° + (RT/F) \log(a_{H^+}) \qquad (Eq. 6.7)$$

where E is the measured potential; $E°$ a constant; and R, T, and F have their usual values. This equation may be simplified to

$$E = E° - 58.2(pH) \qquad (Eq. 6.8)$$

where the potentials are expressed in millivolts. The reference electrode in the chain may be constructed in a number of ways, but the most common

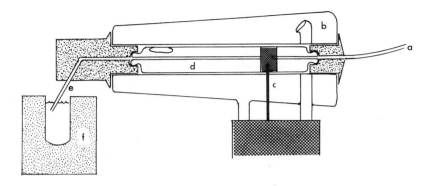

Fig. 6.3 The Radiometer capillary gun pH electrode, showing the sample capillary tubing (a), the temperature jacket (b), the pin contact for the glass electrode (c), the glass electrode insert (d), the connecting capillary (e), and the well-type calomel reference electrode (f).

configuration nowadays is a double-barreled "combination" electrode incorporating a calomel ($Hg/HgCl_2/$ saturated KCl) electrode communicating with the solution to be measured via a porous ceramic plug near the tip of the barrel. The inner glass electrode is sealed, and the outer reference electrode is usually provided with a fill hole for replenishing the KCl. The user should be aware that good electrical contact between the reference cell and the solution is essential, so any dirt that plugs the porous pin will adversely affect the electrode's performance. In the design of the electrodes, the KCl is meant to flow through the porous plug, albeit at a low rate, so prolonged immersion in a small volume of solution may add significant quantities of KCl. For situations in which this is unacceptable, a separate glass electrode must be used with either a double-bridge reference or a reference electrode based upon a different electrolyte.

In many physiological applications, only small sample volumes are available, and so electrodes have been constructed in various ways to accommodate samples down to about 30 μl. One such arrangement is shown in Fig. 6.3; in this electrode the pH-sensitive portion is formed into a small-volume capillary. Contact with the reference electrode is achieved by placing a tube connected to the capillary bore into the KCl well of a calomel reference electrode.

pH electrodes must be calibrated empirically with known buffers, so the precision of the measurement is only as good as the precision of the buffer. Under ordinary circumstances, it is seldom possible to obtain absolute precision greater than ± 0.005 pH units, and even that requires the greatest attention to buffer calibration, temperature control, reference chain conditions, and electrical stability.

The electrical equipment employed for measurement of the pH electrode potential must be capable of making accurate measurements in the millivolt range and must have a very high input impedance (see Chapter 3). Some lower-resistance pH electrodes are now available, but it is best to make sure that the input impedance of the meter used is at least a factor of 10^3 higher than the output impedance of the electrode. Depending on the type of electrode, this means an input impedance of 10^{10} to 10^{15} ohms. Meters used for pH measurement usually make the scaling conversions from millivolt to pH units and are equipped with offset and range controls to set the calibration. Some are also provided with automatic temperature compensation, but one should take the trouble to learn exactly how these "black box" types of controls actually work in order make sure that the corrections are made properly. The meter will always display a number—the trick is in knowing when the number is correct.

6.3.2. pH Indicators

Quite a large number of organic compounds are known which have the interesting property of undergoing a color change as pH changes, and several of these are useful as pH indicators. Some of the more common indicators are listed in Table 6.2, along with information on the pH range of transition and a description of their color changes. The uses of these indicators in bulk solution are mostly obvious, such as for end-point indicators in titration procedures, but some less obvious applications have also been found, particularly in the study of intracellular contents and kidney studies of acidification by kidney tubules.

6.4. MEASUREMENT OF INTRACELLULAR pH

6.4.1. The Weak Acid Method: DMO

The measurement of intracellular pH using the distribution of weak acids and bases dates back to the use of ammonia in the 1870's, but Waddell and Butler's introduction of DMO (5,5-dimethyl-2,4-oxazolidinedione) was a considerable advance because of its non-volatile nature and simpler chemistry. The principle of the method is illustrated in Fig. 6.4, which shows that the un-ionized form of the acid crosses cell membranes relatively freely, whereas the ionized form does not. Since the dissociation equilibrium, given by

$$DMOH \longleftrightarrow DMO^- + H^+ \qquad \qquad \text{(Eq. 6.9)}$$

must be observed both inside and outside the cell, and since DMOH at equilibrium will be equal inside and outside, the relative distribution of

TABLE 6.2
The Properties of Some Common pH Indicators[a]

Name	Transition pH	Color change
Methyl violet	0.0–1.6	Yellow to blue
Malachite green	0.2–1.8	Yellow to blue-green
Thymol blue	1.2–2.8	Red to yellow
Bromophenol blue	3.0–4.6	Yellow to blue
Methyl orange	3.2–4.4	Red to yellow
Ethyl red	4.0–5.8	Clear to red
Alizarin red S	4.6–6.0	Yellow to red
Bromocresol purple	5.2–6.8	Yellow to purple
Brilliant yellow	6.6–7.8	Yellow to orange
Neutral red	6.8–8.0	Red to amber
Cresol red	7.0–8.8	Yellow to red
Phenolphthalein	8.2–9.8	Clear to red
Thymolphthalein	9.4–10.6	Clear to blue
Alizarin yellow R	10.1–12.0	Yellow to red
Clayton yellow	12.2–13.2	Yellow to amber

[a]Modified, with permission, from the *Handbook of Chemistry and Physics.*
Copyright CRC Press, Inc., Boca Raton, Florida.

DMO^- (which is the predominant species) is a function of the pH difference between the inside and the outside of the cell. The intracellular pH can be calculated as

$$pH_i = pK_{DMO} + \log[(DMO_i/DMO_e)(1 + 10^{pH_e - pK_{DMO}}) - 1] \quad (Eq.\ 6.10)$$

Inspection of this rather complex equation reveals that only three measurements need be made: the DMO concentration inside the cell, the DMO concentration outside the cell, and the pH outside the cell. The pK for DMO is dependent upon temperature but lies between 6.1 and 6.3. In the original method, the concentrations of DMO were measured chemically, but the standard technique now employs radioactively labeled DMO, which may be obtained from several companies. An exhaustive review of the use of DMO, as well as other weak acids and bases, for the measurement of intracellular pH has been published recently by Roos and Boron (1981).

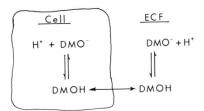

Fig. 6.4 Illustration of the principle of the DMO method for measurement of intracellular pH. (See text for discussion.)

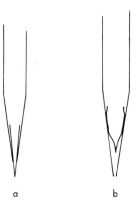

Fig. 6.5 Drawings of (a) the Caldwell type exposed tip pH microelectrode and (b) the Thomas-type recessed tip electrode. In both cases, the inner pH-sensitive glass is heat sealed to the outer tube, which is usually a borosilicate glass with a higher melting point and electrical insulating properties.

6.4.2. pH Microelectrodes

Exposed Tip Electrodes. The first type of pH microelectrode described for intracellular use was constructed as shown in Fig. 6.5a. The outer barrel is made from a non-conductive glass pulled to a fine tip with a standard micro-pipette puller. The inner tip section is drawn from pH-sensitive glass to a very fine tip, which is heat sealed. The inner section is inserted through the outer barrel, heat-sealed to it, and then broken off inside, so that the finished electrode has an open lumen and a short exposed section of pH-sensitive glass. This electrode is inserted into the cell, along with a KCl reference microelectrode, and is employed for pH measurements in the same fashion as a larger electrode (Section 6.3.1). These electrodes have very high resistance, and so a high input impedance pre-amplifier is usually mounted close to the electrode to boost the signal and provide better noise immunity.

Recessed Tip Type. The exposed tip electrode has the disadvantage that it must be inserted into the cell to the entire depth of the exposed section, often 10 μm or more, which precludes its use in small cells. The larger hole made in the cell also increases the problem of electrical leakage around the puncture site. Another type of electrode has been developed by Thomas in which the outer barrel has only a small tip opening (often less than 1 μm) and the pH-sensitive portion is sealed in a recess behind the tip, as in Fig. 6.5b. These electrodes are more difficult to make (see Thomas, 1974, 1978), but allow very shallow penetration of the cell and a small puncture.

Liquid Ion Exchanger Type. More recently, several liquid ion exchangers have been described that are far easier to make than the electrodes described above. They also have the advantage that double-barreled tubing (theta type)

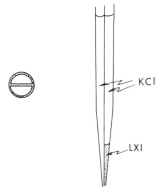

Fig. 6.6 Diagram of a double-barreled liquid ion exchanger pH microelectrode, with an inset showing the cross-section of the theta tubing.

may be used and the reference electrode incorporated into the same piece of tubing. An electrode of this type is shown in Fig. 6.6, along with a cross-section of the tubing used.

6.5. MISCELLANEOUS pH METHODS

Although the methods discussed above are the most common ones in use, other techniques have been developed for specialized applications. One is the use of nuclear magnetic resonance (NMR) spectroscopy for assessing intracellular pH. The method exploits the differences in the spin transitions of ^{31}P between $HPO_4^=$ and $H_2PO_4^-$, which is a function of intracellular pH. The method has the advantages of being non-destructive and capable of detecting rapid transients of pH. Originally the method was only applicable to isolated cells, but more recently larger machines have been employed in studies of organs and whole animals up to the size of humans. There are occasional uncertainties as to just what is being measured, particularly if the intracellular fluid is highly compartmentalized and if different pH regimes predominate in different compartments.

Another method that has occasionally been used is the measurement of fluorescence of certain dyes and indicators that can be introduced into cells. Many of these are in fact weak acids whose fluorescence depends on the amount of the acid remaining undissociated, so any shift in fluorescence is a directly measurable function of pH. One problem with these methods are that the intracellular ionic strength affects the pK of the materials, as do various other intracellular substances, so it is not always clear that the calibration measured *in vitro* is valid within the cell. There are also difficulties in applying the method to anything other than isolated cells.

LITERATURE CITED

Boutilier, R. G., & G. Shelton. 1980. The statistical treatment of hydrogen ion concentration and pH. J. Exp. Biol. 84: 335–339.

Roos, A. & W. F. Boron. 1981. Intracellular pH. Physiol. Rev. 61: 296–434.

Thomas, R. C. 1974. Intracellular pH of snail neurones measured with a new pH-sensitive glass microelectrode. J. Physiol. (Lond.) 238: 159–180.

Thomas, R. C. 1978. Ion-Sensitive Intracellular Microelectrodes: How to Make and Use Them. Academic Press, New York. 110 pp.

Waddell, W. J. & T. C. Butler. 1959. Calculation of intracellular pH from the distribution of 5,5-dimethyl-2,4-oxazolidinedione (DMO). Application to skeletal muscle of the dog. J. Clin. Invest. 38: 720–729.

CHAPTER
7

AMMONIA

7.1. BASIC PROPERTIES

Formula:	NH_3
Molecular weight:	17.03
Boiling point:	$-33.35°C$
Solubility in H_2O:	33.1% w/w at 0°C
pH of 1 N solution:	11.6
Diffusion in H_2O:	1.77×10^{-5} cm² sec⁻¹

Miscellaneous. Corrosive; toxic vapor; infrared absorption bands at 3.00, 6.14, and 10.53 μm; chemically reactive.

7.2. INTRODUCTION

Ammonia is important in physiology because many groups of animals, especially aquatic animals, produce it as the primary end product of nitrogen catabolism. These are the so-called *ammonotelic* animals. Even in groups (such as man) which have other nitrogenous end products, ammonia may be a continuously present minor component, and in certain disease states may reach high concentrations in the blood. Ammonia also plays an important role in the physiology of the kidney in vertebrates. The literature on ammonia production, transport, and excretion has often suffered from the lack of a clear understanding of the physico-chemical behavior of ammonia in dilute

TABLE 7.1
Solubility Values of Ammonia in Water and Fish Plasma[a,b]

Temperature (°C)	Water (L torr^{-1})	Water (mM torr^{-1})	Plasma (mM torr^{-1})
0	1.460	66.09	68.34
10	1.174	53.15	54.95
15	1.050	47.53	49.15
20	0.933	42.24	43.67
25	0.822	37.21	38.48
30	0.719	32.55	33.66

[a]Reproduced, with permission, from Cameron & Heisler, (1983).
[b]The values given are the Bunsen coefficients, with units as stated.

aqueous solutions. Ammonia is particularly tricky because of its behavior as a dissolved gas, as an ion, and as a weak acid in ionic form.

7.3. AMMONIA SOLUBILITY

The solubility of ammonia (i.e., NH_3) in water is extremely high, probably due to the somewhat polar nature of the tetrahedral molecule and the ability to form hydrogen bonds in a fashion similar to that of the water molecule. The polarity is not as great as that of water, however, and liquid ammonia is intermediate in its solvent properties between water and ethyl alcohol. Temperature has an important influence on ammonia solubility, but ionic strength has only a small effect. Oddly, increasing the ionic strength appears to increase the solubility of ammonia slightly, in contrast to most other gases, whose solubility declines at higher ionic strength. The solubility of NH_3 in water and fish plasma is given at various temperatures in Table 7.1.

7.4. THE CHEMISTRY OF AMMONIA IN SOLUTION

Since ammonia is so soluble in water, in the physiological range of concentrations (which rarely exceeds 1 mM L^{-1}) the partial pressure of NH_3 gas in solution is very low, on the order of 10^{-5} torr. In addition to dissolving freely in gaseous form, ammonia reacts with water according to the reaction

$$NH_3 + H_3O^+ = NH_4^+ + H_2O \qquad \text{(Eq. 7.1)}$$

which may also be written as

$$NH_3 + H_2O = NH_4^+ + OH^- \qquad \text{(Eq. 7.2)}$$

which emphasizes the alkaline result. The equilibrium constant is calculated as

$$K' = [NH_3][a_{H^+}]/[NH_4^+] \qquad \text{(Eq. 7.3)}$$

and has a value at 15°C of 3.13×10^{-10}. Thus the reaction in the direction of NH_4^+ + is heavily favored at pH values in the neutral range, NH_3 being favored only at pH values above 9.5. For example, at a pH of 7.5 and a concentration of NH_4^+ of 1 mM L^{-1}, the NH_3 concentration would be only 0.003 mM L^{-1}, or 0.3% of the NH_4^+ concentration. The reaction is very rapid, however, so the two forms always coexist in equilibrium.

The equilibrium constant, K', or its more commonly used negative logarithm, pK', is importantly influenced by both temperature and ionic strength. These effects are large enough to cause significant errors if ignored. As an example, let us say that the pH and total ammonia concentration (NH_3 plus NH_4^+) are known for the blood of an animal and for the water surrounding it. A calculation of the NH_3 partial pressure gradient between the blood and the medium would be considerably in error if there were significant ionic strength differences between the blood and the water. The values for the pK' of ammonia at different temperatures and ionic strengths can be found from the nomogram given as Fig. 7.1.

From the total ammonia concentration and the value for pK', one can calculate the individual values for $[NH_3]$ and $[NH_4^+]$ from the Henderson–Hasselbalch equation:

$$pH = pK' + \log([NH_3]/[NH_4^+]) \qquad \text{(Eq. 7.4)}$$

and since the concentration and partial pressure of NH_3 are related by

$$[NH_3] = \alpha P_{NH_3} \qquad \text{(Eq. 7.5)}$$

the partial pressure can be calculated from (7.4) and (7.5) as

$$P_{NH_3} = (1/\alpha)[NH_4^+]10^{pH - pK'} \qquad \text{(Eq. 7.6)}$$

using values for pK' from the nomogram (Fig. 7.1) and solubility (α) values from Table 7.1.

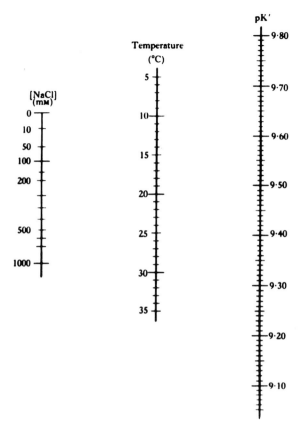

Fig. 7.1 Ammonia pK nomogram. To find the proper value for the pK' of ammonia, follow a straight line from the equivalent NaCl ionic strength on the left through the temperature on the center scale to the value on the pK' scale on the right. (Reprinted with permission from Cameron and Heisler, 1983.)

7.5. MEASUREMENT OF AMMONIA IN GASES

Ammonia is very difficult to measure in the gas phase, since it combines so easily with any water present and has a high tendency to adsorb onto any available surface. In addition to this "sticky" behavior, it corrodes many materials, and is generally a very difficult proposition. The easiest way to analyze the ammonia content of a gas is to trap the NH_3 in a liquid and then analyze it by one of the methods described below. Almost any acidic solution will do, since at pH values below 7, the NH_3 almost completely reacts to NH_4^+ and is effectively trapped in the solution. One simple method would be to absorb the ammonia in a standardized quantity of dilute acid, then back-titrate to obtain the ammonia concentration by difference. If other basic components are present in the gas, this method may not be specific enough, however.

7.6. MEASUREMENT OF AMMONIA IN SOLUTION

7.6.1. Indophenol Method

Ammonia may be made to react with hypochlorite and phenol to form indophenol, which has an intense blue color. Several variations of this method have been described, but they are all fairly similar and depend upon colorimetric determination of the ammonia as a function of the intensity of the blue color (Solorzano, 1969). Known standards are carried through the same analytical procedure for calibration. For physiological studies, this method is good for dilute solutions that are relatively protein free and can usually be used for fresh or salt water and for urine samples. There are a number of possible interferences, however, particularly from protein and free amino acids. The method requires a high pH, and hydrolysis of NH_3 from these compounds may occur. Ultraviolet light interferes slightly with the reaction, so sunlight or high intensity fluorescent lighting should be avoided during color development. Another problem with the method is the toxicity and unpleasant odor of phenol; a salicylate modification of the method has been described that avoids this problem.

These wet chemical methods adapt themselves fairly well to automated analysis devices, such as the Techni-Con AutoAnalyzer. The actual procedure involves merely sequential addition and mixing of three reagents, an appropriate delay loop, and a colorimetric detector. When large numbers of samples are to be analyzed, an automated system for ammonia can be a great time-saver.

7.6.2. Enzymatic Methods

For blood and other biological fluids that contain protein or other nitrogenous compounds that may interfere with the indophenol method, a specific enzymatic assay is now available in commercial kit form from several manufacturers (Sigma, Boehringer, etc.). The assay is based on the reaction

$$\alpha\text{-Ketoglutarate} + NH_4^+ + NADH \rightarrow \text{glutamate} + NAD \qquad (Eq. 7.7)$$

which is catalyzed by L-glutamate dehydrogenase. The quantity of ammonia in an unknown sample is directly proportional to the reduction in the absorbance of NADH at 340 nm. This reaction is specific for the ammonium ion and is carried out under conditions that do not lead to hydrolysis of ammonia from other compounds that may be present. The commercial kits are usually designed for a 3 ml total assay volume and sample volumes around 200 μl, but can easily be scaled down to 1 ml or smaller, with corresponding reductions in the sample needed.

7.6.3. Ammonia Electrodes

In recent years commercial electrodes have become available for the measurement of ammonia in solutions, and with certain limitations these have proved quite useful. Information on some types is difficult to come by, since it is proprietary information, but the most common types appear to be in essence pH electrodes coupled with a membrane and electrolyte system in a fashion very similar to that of the CO_2 electrode (Section 5.1.1). The electrode is contained in a sleeve filled with a buffer/electrolyte, separated from the solution to be measured by a special NH_3-permeable membrane (Fig. 7.2). In order to measure the total ammonia content of an unknown, since most of the ammonia is present as NH_4^+, the sample must be brought to a pH sufficiently alkaline to convert nearly all of the ammonia to NH_3. Usually the electrodes are employed in solutions brought to about pH 12. When the external pH is made sufficiently alkaline, the NH_3 equilibrates across the membrane with the internal electrolyte. The internal electrolyte is a solution of NH_4Cl sufficiently concentrated so that its concentration may be considered constant, and (7.3) reduces to

$$K' = [NH_3][a_{H^+}]$$ (Eq. 7.8)

and so the pH of the internal electrolyte will be proportional to the ammonia content of the sample.

Some of the problems in using ammonia electrodes are common to CO_2 electrodes, particularly the necessity for careful temperature control. This is actually more critical for ammonia than for CO_2, since the physicochemical characteristics of ammonia are more temperature sensitive. Another common problem is the electrolyte capacitance, or memory effect. The electrolyte volume in ammonia electrodes tends to be large, so if the sample to be measured is very small, some error may result. A further disadvantage is the requirement for bringing the sample to pH 12 or so, which means that for most materials it is a destructive method. When this is done with some

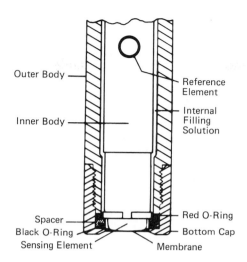

Outer Body

Reference
Element

Internal
Filling
Solution

Inner Body

Spacer

Red O-Ring

Black O-Ring

Bottom Cap

Sensing Element

Membrane

Fig. 7.2   Cross-section of an ammonia electrode. (Courtesy of Orion Research, Inc.)

solutions, precipitation may occur, fouling the membrane and reducing the
stability and accuracy of the method. In sea water there is considerable
precipitation, mostly of carbonates and sulfates, which do not appear to
interfere greatly, but in blood, protein precipitation on the membrane is fatal.

The ammonia electrode can be particularly useful when continuous
measurements are required over long periods of time. An automated,
computer-controlled system for ammonia measurement is depicted in Fig.
7.3. In this system, the ammonia electrode is mounted in a small-volume
chamber which is equipped with two inlets, an outlet, a small stirring bar,
and a temperature jacket. The two inlet tubes are connected to a roller pump
which delivers sample through one channel and 10 N NaOH through the
other. The sample-to-base ratio is about 6 to 1, which brings the sample
to the proper pH when mixed in the sample chamber. The output of the meter
connected to the ammonia electrode is converted to digital form under
computer control, and the computer also controls two three-way valves which
allow periodic calibration of the electrode with standard ammonia solutions.
The clock in the computer times the sequence of calibration and sampling,
and other program routines decide when the electrode has stabilized at each
calibration and sample point, as well as performing any other desired
calculations and display functions. The system shown has been used for
measuring ammonia excretion rates of animals in closed-system recirculating
aquaria, and with the addition of one more three-way valve it could be used
for a flow-through system, switching between inflow and outflow streams.

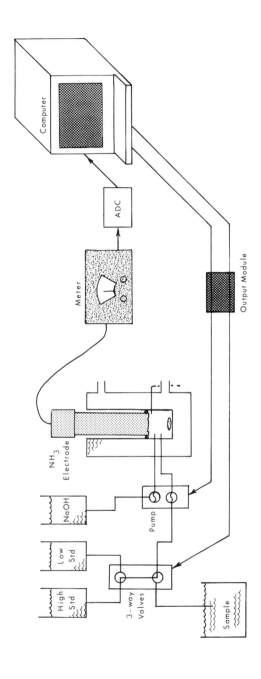

Fig. 7.3 A functional block diagram of a computer controlled ammonia measurement system. For a complete description, see text.

7.6.4. Microdiffusion Method

An older but still good method for measurement of ammonia is the so-called microdiffusion method (Conway, 1957). The principle of analysis is to convert the total ammonia in the sample to NH_3 by alkalinization, then allow the NH_3 to diffuse through a gas phase into a second acidic solution which acts as an ammonia trap. If the amount of acid in the second solution is measured precisely beforehand and is back-titrated after complete diffusion, then the ammonia content of the original sample may be calculated by difference. The reaction is carried out in a microdiffusion vessel such as the one shown in Fig. 5.7. The vessel is pre-charged with alkali in the center well and the standardized acid in the outer ring. After addition of the sample to the center well, a flat lid is sealed in place with grease and the diffusion is allowed to go to completion, often overnight. The method has the advantages of simplicity and very low cost compared to some other methods in use.

7.7. AMMONIA CONTAMINATION

Since the total concentrations of ammonia are often very low in samples of physiological interest, the problem of ammonia contamination of both reagents and distilled water is persistent and serious. In work with natural waters the problem is even more acute, since the concentrations of interest may be as low as $0.01 \mu M L^{-1}$. Ordinary lab distilled water is generally not a good source for ammonia analysis, since the ammonia will distill over with the water at high temperatures and at the often high pH values in stills. Commercial ion-exchange units found in many laboratories are often worse, since either ammonia or organic compounds that release ammonia upon decomposition are frequently leached out of the resins. One good way to produce an ammonia-free water is to mix a good grade of distilled water with a strongly acidic cation-exchange resin, such as Dowex 50W-Hydrogen. After a few minutes of mixing, the resin may be allowed to settle or separated by filtration, and the resulting ammonia-free water kept tightly stoppered. If filtration is employed, the filter should be carefully checked for ammonia contribution. Storage should be in glass rather than (permeable) plastic, and this ammonia-free water should then be used for reagents, blanks, and glassware rinse.

LITERATURE CITED

Cameron, J. N., & N. Heisler. 1983. Studies of ammonia in the rainbow trout: physico-chemical parameters, acid–base behaviour and respiratory clearance. J. Exp. Biol. 105: 107–125.
Conway, E. J. 1957. Microdiffusion analysis and volumetric error. Crosby Lockwood, London.

Solorzano, L. 1969. Determination of ammonia in natural waters by the phenolhypochlorite
    method. Limnol. Oceanogr. 14: 799–801.

SUGGESTED FURTHER READING

American Public Health Association. 1976. Standard Methods for the Examination of Water
and Wastewater. New York.

CHAPTER
8

MISCELLANEOUS GASES

8.1. HYDROGEN

8.1.1. Basic Properties

Formula:	H_2
Atomic number:	1
Atomic weight:	1.008
Valences:	$+1$
Melting point:	$-259.2°C$
Boiling point:	$-252.8°C$
Density, 0°C:	0.0899 g L^{-1}

Miscellaneous.   Odorless, colorless gas; diatomic molecule, H_2; stable isotope ^2H, deuterium, with natural abundance of 0.02%; unstable isotope ^3H, tritium; free gas rare in nature, about 1 ppm of atmosphere; 10.8% by weight of the oceans.

8.1.2. Chemical Properties

At ambient temperature, hydrogen is relatively unreactive, but at high temperatures or in the presence of catalysts like platinum, it is highly reactive. The reaction with oxygen to form water, for example, does not proceed at a measurable rate at room temperature, but when initiated by catalysis, a spark, or a flame, it may proceed with explosive violence. Hydrogen is a powerful reducing agent, reacting not only with various metals but with unsaturated organic compounds of many kinds. It may be produced in the decomposition of organic matter by certain classes of bacteria or blue-green algae.

Preparation of hydrogen gas in the laboratory may be easily carried out by the reaction

$$Zn + 2\,HCl \rightarrow ZnCl_2 + H_2 \tag{Eq. 8.1}$$

Industrially, most commercial hydrogen is prepared from petroleum, and its principal use is in the manufacture of ammonia and other chemicals.

8.1.3. Analytical Methods

In spite of the large number of possible reactions undergone by hydrogen, there is no very satisfactory chemical method for its analysis. The 19th century chemists employed the reaction with oxygen, measuring the reduction of the gas volume after formation and condensation of the water vapor. A less explosive oxidation may be carried out by passing the gas over hot copper oxide, but this method is still rather cumbersome and lengthy. Most practical analytical methods depend upon the physical properties of hydrogen gas, particularly its density and thermal conductivity. Mass spectrometry is a very "clean" method for hydrogen analysis, since there are no other gases that overlap its mass range, and the sensitivity can be quite high. This method is particularly suitable for complex gas mixtures whose other components may not be known.

Gas chromatography is also quite suitable, although since hydrogen and helium both have high thermal conductivity, nitrogen may be used as a carrier gas. The sign of the detector signal is simply reversed, since hydrogen has much greater thermal conductivity than the carrier gas.

The polarographic oxygen electrode may also be used for hydrogen analysis by simply reversing the polarizing voltage. The cathodic reaction then depends upon the H^+ concentration, and the electrode current may be converted to a convenient form for measurement by the same means as for oxygen analysis (see Chapter 4).

8.1.4. Physiological Applications

Hydrogen is employed relatively rarely in physiology. Some applications depend upon its reaction with oxygen; for example, to ensure that a carrier gas (such as nitrogen) is oxygen free, sometimes small percentages of hydrogen are added to it and the gas is passed over platinum catalyst beads. Any traces of oxygen remaining in the gas stream will react with the hydrogen to form water, which can then be removed in a water trap. This method is used in some galvanic oxygen analyzers and might be used for de-oxygenating nitrogen in studies of blood pigments.

Since hydrogen is relatively soluble in aqueous solutions and is highly diffusible in air and water, it is sometimes used as a tracer for diffusive exchange. For example, a single breath of a hydrogen mixture may be followed, both in the expired air and in the blood, by a combination of mass spectrometry and polarography. The single breath of hydrogen then acts as a marker, and the time course of washout in the blood and gas from the lungs provides data about the flow rates and exchange conditions.

8.2. CARBON MONOXIDE

8.2.1. Basic Properties

Formula	CO
Molecular weight	28.0
Melting point	$-205.1\,°C$
Boiling point	$-190.0\,°C$
Density, $0\,°C$	1.250 g L^{-1}

Miscellaneous. Colorless, odorless, tasteless gas; stable at room temperature; prepared from coke or coal and steam at high temperature; highly poisonous.

8.2.2. Chemical Properties

Carbon monoxide is moderately reactive with a number of classes of compounds, some of which produce toxic products. CO and chlorine in the presence of UV light combine to produce the toxic gas phosgene, $COCl_2$. CO also reacts with various halides and metals to form carbonyl compounds. The most significant reaction for physiology is the binding of CO to the heme group in hemoglobin. Although this is a reversible reaction, like the oxygenation of hemoglobin, the affinity of the heme group for CO is an order of magnitude greater than for O_2. Inhalation of CO at a concentration of only 0.09% will lead to nausea and headache, and slightly greater concentrations will lead to death. The cause of death is anoxia, since the hemoglobin bound to CO cannot function as an oxygen carrier. CO will eventually be cleared from the body, but 24 hr or more may be required.

Carbon monoxide may be readily absorbed by a concentrated solution of cuprous chloride in HCl or in NH_4OH. It also decomposes at high temperature by the reaction

$$2 CO \rightarrow C + CO_2 \qquad\qquad (Eq. 8.2)$$

and this reaction may be catalyzed to proceed at room temperature by Hopcalite, a mixture of the oxides of manganese and copper. CO may be readily prepared in the laboratory by heating $CaCO_3$ and Zn powder, but of course should be handled with great care, due to its highly toxic properties.

8.2.3. Analytical Methods

Since CO has quite a low thermal conductivity, it may be conveniently analyzed by gas chromatography. This method is extremely sensitive and will provide reliable analyses in the presence of complex mixtures of other gases. Mass spectrometry would appear to be a logical choice, but the mass of CO (28) happens to coincide with the mass of N_2, limiting the applicability of this method. The Scholander and Van Slyke apparati were also employed for analysis of CO, using various chemical absorbents and measuring the reduction of the gas volume manometrically (see Chapters 4 and 5; Harrington & Van Slyke, 1924; Scholander, 1947; Van Slyke & Plazin, 1961).

8.2.4. Physiological Applications

Carbon monoxide has been used primarily for the study of diffusing capacities in the mammalian lung, a method which was first described by Krogh in 1909. CO is well suited for this purpose, since the binding reaction with hemoglobin is so fast and complete that the movement of CO from lung gas to blood is effectively a one-way diffusive process, uncomplicated by back-diffusion. In practice, the subject either re-breathes a closed volume of gas containing CO at about 0.5% concentration or takes a large single breath and holds it. The rate of disappearance of the CO will follow, more or less, a simple exponential, and from that the diffusing capacity of the lung, DL_{CO}, may be calculated (Forster, 1964). The fine points of this technique are still being debated, but it appears to have a certain usefulness in studies of lung function.

CO has found some secondary application in blood studies, where it may occasionally be convenient to de-activate the hemoglobin by exposing it to CO. Graded exposure to CO may be used to produce hypoxia of varying degrees, and it has been used in fish to study the effects of completely blocking the hemoglobin function (Nicloux, 1923; Anthony, 1961).

8.3. HELIUM

8.3.1. Basic Properties

Formula	He
Atomic number	2
Atomic weight	4.0026
Valences	0
Melting Point	$-272.1°C$ (1.1 K)
Boiling Point	$-268.94°C$ (4.22 K)
Density, 0°C	0.1785

Miscellaneous. Colorless, odorless, tasteless gas; second lightest gas; naturally occurring isotope ^{3}He extremely rare; major source: natural gas fields in the United States.

8.3.2. Chemical Properties

Helium is the lightest element of the so-called group zero of the periodic table, and has extremely high chemical stability due to the configuration of its electron shells: it has only the first shell, which is completely filled by two electrons. A few compounds of helium have been produced, but only under highly artificial laboratory conditions. The density and viscosity of helium gas are very low, and the thermal conductivity and specific heat are very high. This makes helium a good choice as a carrier gas for gas chromatography, since every gas except hydrogen has a considerably lower thermal conductivity. Due also to its orbital configuration, helium has the lowest condensation temperature of any known substance and is finding increasing application in electronics, where liquid helium is used for inducing superconductivity.

The natural supply of helium has become a matter of considerable concern. Natural gas from certain fields in Texas, Oklahoma, and Kansas contains as much as 1% helium, compared to its abundance in the atmosphere of only 0.0005%. Although those wells that are highest in helium are mined for it, many other natural gas sources that contain amounts significantly above the atmospheric concentration are not scavenged, and the He is eventually released to the atmosphere. When the high-concentration wells are exhausted, the energy costs of He extraction from the atmosphere will probably lead to a shortage and at least an order of magnitude increase in cost. Argon may be substituted in many uses for He, such as in welding, and since it is far more abundant (0.93% of the atmosphere), it is far cheaper.

8.3.3. Analytical Methods

Gas chromatography and mass spectrometry are the two principal methods of He analysis, although an alternate carrier gas must be used with the former method. Manometric analysis is at least theoretically possible by absorbing everything else from a gas sample and measuring the volume remaining. This is not a practical method, however. The presence of He can also be detected by its strong yellow emission line, which was the property which first led to its discovery in the sun's spectrum in 1868 and to its discovery on earth almost 30 years later.

8.3.4. Physiological Applications

The main physiological use of helium, aside from its use as a carrier gas in various instruments, is as an inert marker gas for the study of exchange and transport by the respiratory system. It has some of the advantages of CO for determining diffusing capacity (see above), plus the added advantage of being non-toxic. The use of CO is limited to very short periods and low concentrations, whereas the entire body can, if necessary, be saturated with He. The use of He in special diving gas mixes does just that, relying on the lower solubility and higher diffusion coefficient of He (compared to N_2) in water and body fluids to reduce the hazards of the bends.

8.4. NITROGEN

8.4.1. Basic Properties

Formula	N_2
Atomic number	7
Atomic weight	14.0067
Valences	$+5$ to -3
Melting point	$-210.0°C$
Boiling point	$-195.8°C$
Density, $0°C$	1.251

Miscellaneous.   Most abundant component of the atmosphere; relatively inert; colorless, odorless, tasteless gas; asphyxiant in high concentrations;

prepared by liquefaction of air; stable isotope ^{15}N is 0.37% in natural abundance.

8.4.2. Chemical Properties

From the standpoint of animal physiology, nitrogen may be considered unreactive, although it does undergo various reactions in plant metabolism. Nitrogen has an unusually large range of valence states, illustrated by the oxides N_2O_5 ($+5$), NO_2 ($+4$), N_2O_3 ($+3$), and NO ($+2$); and by the hydrogen compounds HN_3 (azides, $-\frac{1}{3}$), NH_2OH (-1), NH_2NH_2 (-2), and NH_3 (-3). A number of these compounds have biological significance, either biochemically or physiologically. Nitrous oxide, for example, is a commonly used analgesic/anesthetic. A detailed account of the chemistry of each of these compounds, however, is outside the scope of this book and may be found in various chemistry texts.

8.4.3. Analytical Methods

In a respiratory gas, the volume percent of nitrogen may often be obtained simply as the residue remaining after extraction of oxygen and carbon dioxide. Argon is the only other component of the atmosphere that would cause a significant error. Gas chromatography is also highly suitable for N_2 analysis with a variety of column materials and flow conditions. Mass spectrometry is also well suited, provided CO is not a major component of the gas to be measured (see above).

Since nitrogen has no useful radioactive isotopes (all have half-lives of less than 1 min), tracer work must be done using the naturally occurring stable isotope ^{15}N, and analysis of the tracer depends heavily on accurate isotope ratio mass spectrometry. Even so, the sensitivity is far less than what is routinely achieved with radiotracers (see Chapters 14 and 15), and studies of nitrogen metabolism have been correspondingly hampered.

8.4.4. Physiological Applications

Nitrogen is encountered principally as an inert "filler" for the atmospheric and respiratory gases, and one does not normally think of its applications. Since it is a common and inexpensive gas, it is usually the one chosen for de-oxygenation in various procedures, such as equilibration of blood for dissociation curves, as a carrier gas in some oxygen and CO_2 analyzers, etc. Other gases would do as well (helium, for example), but they are much more costly.

In respiratory physiology, one important consideration of the behavior of nitrogen is in diving. Breathing of air under pressure for prolonged periods causes saturation of body fluids and tissues with nitrogen. When the diver returns to ambient pressure, the solubility coefficient of nitrogen is exceeded, and bubbles may form from the supersaturated blood and tissues. Formation of bubbles leads to "bends," a very painful condition which may be permanently crippling or fatal, depending upon the degree of severity. The advantages of helium (see above) are lower solubility and a much more rapid diffusion rate, which reduces the time required for return to normal pressures.

LITERATURE CITED

Anthony, E. H. 1961. Survival of goldfish in the presence of carbon monoxide. J. Exp. Biol. 38: 109–125.

Forster, R. E. 1964. Diffusion of gases. *In* Handbook of Physiology, Section 3: Respiration. W. O. Fenn & H. Rahn, eds. Amer. Physiol. Soc., Washington, D.C. Vol. I, pp. 839–872.

Harrington, C. R., & D. D. Van Slyke. 1924. The determination of gases in blood and other solutions by vacuum extraction and manometric measurement. II. Modifications for CO measurements. J. Biol. Chem. 61: 575–584.

Krogh, A., & M. Krogh. 1909. Rate of diffusion of CO into lungs of man. Scand. Arch. Physiol. 23: 236–247.

Nicloux, M. 1923. Action de l'oxyde de carbone sur les poissons et capacité respiratoire du sang de ces animaux. C. R. Seances Soc. Biol. 89: 1328.

Scholander, P. F. 1947. Analyzer for accurate estimation of respiratory gases in one-half cubic centimeter samples. J. Biol. Chem. 167: 235–250.

Van Slyke, D. D., & J. Plazin. 1961. Manometric Analyses. Williams & Wilkins, Baltimore. 89 pp.

Weast, R. C. (Ed.) 1970. Handbook of Chemistry and Physics. CRC Press, Cleveland, Ohio.

CHAPTER
9

CONTROLLING THE GASEOUS ENVIRONMENT

9.1. INTRODUCTION

In several previous chapters, various means for measuring the composition of gas mixtures or for quantitative analysis of specific gases have been discussed. In many physiological applications, however, it is necessary to control the gas composition of the environment. Some aspects of this general problem are intuitively obvious, but there are others which are neither obvious nor always easy to solve. The following sections address some of the more common applications and various suitable techniques.

9.2. CONTROLLING HUMIDITY (WATER VAPOR)

Although water is not normally thought of as a gas, it is nearly always present in any gas mixture. Though minor in terms of its percentage composition of most gases, it is nonetheless an important physiological parameter. The amount of water in a gas is usually expressed as the *relative humidity*, which is calculated as the amount of water in a given volume, divided by the amount which could be held at saturation, multiplied by 100 to obtain a percentage. The actual amount of water vapor contained in air, for example, is 17 g m^{-3} at 20°C and 30 g m^{-3} at 30°C, so a water content of 10 g m^{-3} would represent a relative humidity of 33% at 30°C, and of 59% at 20°C.

9.2.1. Desiccation

The simplest example of humidity control is desiccation, i.e., total removal of water vapor. This can be accomplished in a variety of ways, but the most common method is to pass the gas of interest through a tubular trap containing one of the commercially available drying agents, such as Drierite or silica gel. These drying agents are normally salts that are extremely *hygroscopic*, i.e., they have a strong tendency to absorb water vapor. Anhydrous $CaCl_2$ is one of the commoner salts used, and is often combined with an indicator to show when the material is exhausted. The user should be aware that water removal is not complete using these agents; $CaCl_2$ will reduce the water content to about 1.5 mg L^{-1} of air, a relative humidity of about 9%, whereas anhydrous $CaSO_4$ will reduce the relative humidity to only 0.03%. The most effective agent is P_2O_5, which will remove water to undetectable limits, but it is rather dangerous to handle and so is used only for special purposes. Both Drierite and silica gel may be regenerated by heating to a point sufficient to drive off the absorbed moisture.

Another common method of desiccation is to pass the gas stream through a cold trap so that the water vapor is frozen on the walls of the trap. A slurry of dry ice and either acetone, methanol, or isopropyl alcohol is a convenient bath for the cold trap, although mechanical refrigerating devices may also be used. This method is not as efficient as chemical trapping, so care must be taken to provide sufficient contact time and surface area for the water vapor to be completely removed.

9.2.2. The Solution Method

The affinity of many salts for water may be exploited in another way to control the relative humidity at values greater than zero. In a saturated solution with excess crystalline material present, water vapor in a gas phase over the solution will reach a stable equilibrium. For example, imagine sealing a saturated LiCl solution into a closed vessel with moist air. The affinity of the saturated salt solution for water is such that the vapor pressure of the solution is appreciably reduced over pure water. Since the (partial) pressure of water vapor in both liquid and gas phases will tend to equilibrate (see Section 2.4), there is at first a net movement of water into the solution. If the solution is stirred, the addition of water vapor causes it to fall below saturation, but some of the excess solid LiCl will dissolve, restoring complete saturation. This process continues until the water vapor in the gas phase falls to a point just equal to the vapor pressure of the solution. So long as no heterogeneity develops in the solution and the solid LiCl is not exhausted, the water vapor content (humidity) of the gas phase above it will be kept

TABLE 9.1
The Relative Humidity of Air Equilibrated with Saturated Salt Solutions[a,b]

Salt	% humidity
$H_3PO_4 \cdot \frac{1}{2}H_2O$	9
$LiCl \cdot H_2O$	15
$KC_2H_3O_2$	20
$CaCl_2 \cdot 6H_2O$	32.3
$K_2CO_3 \cdot 2H_2O$	44
$Na_2Cr_2O_7 \cdot 2H_2O$	52
$NaBr \cdot 2H_2O$	58
$NaNO_2$	66
NH_4Cl and KNO_3	72.6
NH_4Cl	79.5
$(NH_4)_2SO_4$	81
K_2CrO_4	88
$Na_2CO_3 \cdot 10H_2O$	92
$Na_2SO_3 \cdot 7H_2O$	95
$CuSO_4 \cdot 5H_2O$	98

[a]Data from the *Handbook of Chemistry and Physics*.
[b]Data are for 20°C unless otherwise stated.

constant at about 15%. A wide range of relative humidity values may be obtained with various salt solutions, as shown in Table 9.1.

Another relatively simple method which will produce a continuous range of relative humidity values is to use solutions of H_2SO_4 of varying density (concentration). Relative humidity values ranging from 100.0 to 3.2% may be obtained in this way, as shown in Table 9.2.

TABLE 9.2
Constant Relative Humidity in Air Equilibrated with Sulfuric Acid of Different Densities at 20°C[a]

Density	Relative humidity	Density	Relative humidity
1.00	100.0	1.30	58.3
1.05	97.5	1.35	47.2
1.10	93.9	1.40	37.1
1.15	88.8	1.50	18.8
1.20	80.5	1.60	8.5
1.25	70.4	1.70	3.2

[a]A specific gravity of 1.20 corresponds to about 28% H_2SO_4 by weight.

A major disadvantage of these methods, of course, is that they are easily applicable only to small volumes of gas. For flowing streams or large gas volumes, other methods must be found. Simple humidification may be accomplished by bubbling the gas stream through water, though it is easy to produce either more or less than 100% saturation, particularly if the temperature of the water is slightly different from that of the gas. This may be very important, for example when equilibrating small liquid volumes with gases. A slightly undersaturated gas will evaporate the sample, and a slightly supersaturated gas will dilute it by condensation. A better method, particularly for low flows, is to bring the gas into contact with a wick saturated with liquid of the same temperature and ionic strength as the liquid sample being equilibrated.

9.2.3. Mechanical Systems

For large volumes of gas, such as those of an entire room, mechanical means must be employed for humidity control. These might include evaporative humidifiers for increasing it or refrigerator/condenser systems for decreasing it, much like those we employ in buildings. The accuracy of control will depend on the adequacy of mixing in the room air and the accuracy of the sensor and controller systems employed.

9.3. MIXING GASES

9.3.1. Cylinder Mixtures

The simplest method for obtaining gas mixtures of any given composition is to purchase them in cylinders from commercial gas suppliers. Most dealers have a standard range of the most common physiological gases, particularly those with clinical uses, and will make up custom mixtures for a reasonable fee. Based upon considerable experience, however, the reader is strongly advised either to conduct his own analysis of the mixture supplied or to deal only with a supplier whose analyses he has come to trust (by checking!). This policy is consistent with the general rule of taking nothing on faith, but the thought of having a whole series of experiments wasted because the gas mixture was not correct is enough to give anyone nightmares. The same advice also holds, incidentally, for pure gases, especially if the best grades are not purchased. It is not uncommon for nitrogen cylinders to have a small amount of oxygen in them, and this can be disastrous for some kinds of work.

The cylinder mixture strategy becomes unsatisfactory if a range of mixtures is required. The laboratory can quickly fill up with partly full gas cylinders, and the demurrage charges pile up. For many applications, a variable mixing system is required.

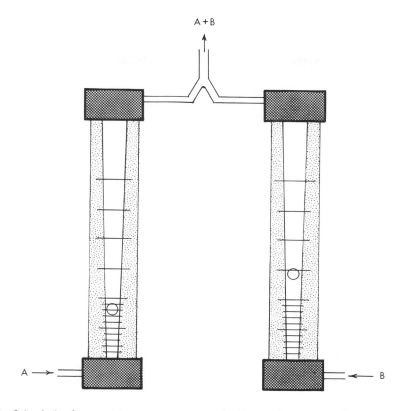

Fig. 9.1 A simple gas mixing system constructed with two flowmeters and a common outlet tube.

9.3.2. Mixing Manifolds

Mechanical mixing manifolds of various kinds can either be purchased or constructed *ad hoc* for many requirements. A simple one might consist merely of two valves and flowmeters connected to a common output tube, as in Fig. 9.1. Provided the gas pressures are well regulated and the gases are clean, allowing no dirt to change the flow valve or meter resistances, these simple arrangements will suffice in many cases. One should be aware, however, that the flowmeters will be calibrated for a particular gas, often air, and that apparently small differences in the viscosity of different gases will cause a predictable error in the flow readings. The values for viscosity of some common gases are given in Table 9.3, as is the relative flow rate, an indication of the error in viscosity-based flowmeters. Temperature also has an effect on the registered flow, since viscosity drops as temperature increases. This, plus the inherent error in the flowmeters, limits these simple manifolds to applications in which the gas mix need only be controlled to within a few percentage points.

TABLE 9.3
Viscosity (η) and Relative Flow Rate of
Some Common Gases at 20°C[a,b]

Gas	η	Relative flow
Air	170.8	1.00
N_2	167.4	1.02
O_2	192.6	0.89
CO_2	138.0	1.24
CO	166.5	1.03
H_2	85.0	2.01
He	188.7	0.91
NH_3	94.4	1.81

[a]Viscosity from Radford (1964) with permission.
[b]The relative flow rate is proportional to the reciprocal of viscosity with air taken as 1.00.

Recently, a simple manifold system was described (as shown in Fig. 9.2) that can provide an accurate mixture of minor components into a major component, as for example in mixing low concentrations of CO_2 into air (Rodeau & Malan, 1981). This system is easily constructed but should probably have a good filter in the gas line to prevent changes in the resistance of the capillary tube. This system would have to be calibrated at each temperature and for each gas used.

9.3.3. Gas Mixing Pumps

An accurate and versatile mixing pump for gases is manufactured in West Germany by Wosthoff (Fig. 9.3). It depends on very accurately machined pistons and gears to mix gases in set proportions. By changing gear ratios, one can vary the mixtures achieved through a range of 1 to 100% in 1% steps. Other models are available for special mixture ranges or for corrosive gases. Two or more pumps may be connected in series to extend the range below 1% or to obtain fractional percentages. The major drawback of these pumps is cost, which is currently around $9000 for the model shown, but if a wide range of gas mixtures is routinely required, the investment may be justified. There are two pumps in the author's laboratory that have been used heavily for 12 years. Another limitation is that only a single flow rate is possible, about 36 L hr^{-1}, and the pumps must be specially equipped to maintain accurate output against any significant back-pressure on the outlet.

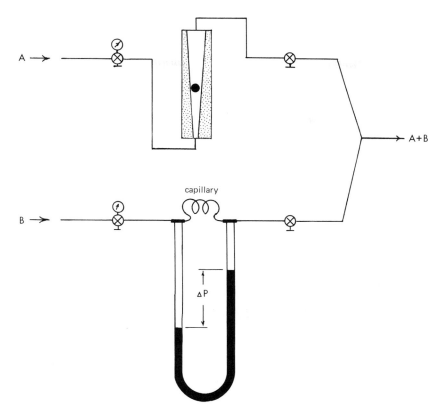

Fig. 9.2 An inexpensive manometer-type flow mixing system for adding a minor component (B) to a major one (A). Flow of the minor component through the capillary resistance is controlled and measured by the pressure gradient (ΔP) across the water manometer. (From Rodeau & Malan, 1981.)

9.3.4. Mass Flowmeters

The limitations on flow range and back-pressure may be avoided by employing one of the commercial flowmetering devices based on thermal conductivity, the so-called mass flowmeters. They employ a flow sensor which is heated and exposed to the flowing gas stream. The rate of cooling of the sensor element is proportional to the thermal conductivity of the gas and its mass flow rate, i.e., the flow rate in grams or moles per unit time. These have the advantage of being virtually temperature insensitive and avoid the temperature corrections necessary with volume-based flowmeters. That is, as the temperature of the gas rises and its density falls, a proportionately

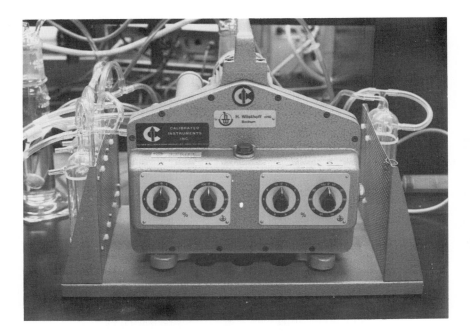

Fig. 9.3 A two-component gas mixing pump manufactured in West Germany by Wosthoff.

larger flow rate is required to achieve the same cooling of the sensor; the increase in volume flow is just that which is necessary to maintain the same mass flow of the expanded gas.

Such systems are offered by several companies and are usually calibrated for the thermal conductivity of a particular gas (Table 4.1; Radford, 1964). Often the flow rate sensor is coupled to a control value to provide both a measurement and a control system. The stated accuracy of one such sensor/controller module (Tylan Corp.) is $\pm 1.0\%$ of the full scale flow range, with a repeatability of $\pm 0.2\%$. So, for two- or three-gas mixtures, the worst case limits will be the product of the individual errors.

9.4. GAS WASHOUT AND CHANGEOVER

It seems obvious that if different gas environments are part of an experimental routine, the gas environment must at some point be changed. The usual procedure is to begin the flow of the new gas, wait some appropriate time, and resume whatever measurements are being made. What is not so obvious is how long it may take for complete (i.e., 99% or more)

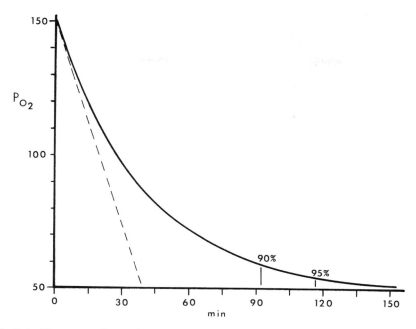

Fig. 9.4 The output of a computer simulation of the time course of oxygen in a chamber after a step change in concentration at time zero. The chamber volume is 40 times the flow per minute. See Appendix 4 for details of the calculation.

changeover from one gas to another. It is really a straightforward application of the equation for exponential decay

$$f(t) = f(0)e^{-kt} \qquad \text{(Eq. 9.1)}$$

where $f(t)$ is the value of concentration at time t and k a rate constant determined by the flow and volume. Simple models of a given experimental apparatus may be analyzed by using an algebraic approximation method to solve (9.1), and the resulting program can be run on any microcomputer, and many hand-held programmable calculators. A sample program is given in Appendix 4, and the solution to a particular problem is shown graphically in Fig. 9.4. In this case, 99% approach to the new concentration is not reached until 116 min, even though the gas flow was equal to the chamber volume each 40 min.

9.5. CONTROLLING GASES IN SOLUTIONS

For a closed volume of water, the easiest method for controlling the dissolved gases in solution is to bubble a gas of the desired composition through it using an air stone or similar diffuser. With small bubbles and a good gas flow rate, the equilibration is reasonably efficient. One should be

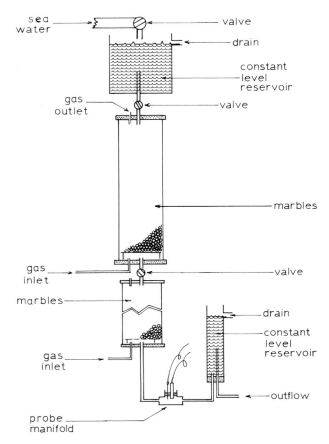

Fig. 9.5 Diagram of a system for control of the partial pressure of O_2 in a running water supply. The system incorporates constant level reservoirs for flow control and a manifold for an oxygen meter for continuous monitoring. (Modified from Fry, 1951.)

aware, however, that bubbles in water have a higher gas pressure than atmospheric because of surface tension in the bubble. A sample of water thoroughly equilibrated with fine bubbles, therefore, will have a higher partial pressure of dissolved gases than the equilibrating gas. Since equilibrium will also be approached at a rate dependent upon the partial pressure difference, the same kinetics apply as for gas changeover (see eq. 9.4). That is, 99% equilibration may take a long time and should be checked carefully by measurements of the dissolved gases.

If rapid changes of dissolved gases are required, almost the only method is to prepare another volume of water ahead of time and rapidly exchange the volumes. In the example of Section 9.4, instantaneous mixing of the complete chamber volume was assumed in modeling gas changeover. For aqueous solutions, the lack of mixing during changeover may be used to advantage by arranging the inflow and outflow in such a way that there is little mixing. This helps to bring the time for complete changeover closer to the no-mixing model, where the time required is simply the volume divided by the flow rate.

For continuous flow systems, a simple and inexpensive stripping column was described by Fry in 1951 and is shown in Fig. 9.5. In this apparatus, the water flows downward over an inert substrate material (e.g., glass beads or marbles) that provides a large surface area for diffusive exchange. The gas flows upward through the wetted beads, providing a counter-current arrangement. At high flow rates the exchange (equilibration) between gas and liquid may not be complete, but so long as the flow rates of both gas and liquid are maintained, the effluent dissolved gas concentrations will be in a steady state. This apparatus can be adapted for small or large volume applications, and two or more columns may be connected in series. For de-oxygenating water, I have found that two columns in series are needed to get less than 10% saturation at flows of 1 to 2 L min^{-1} in a column of about 15 × 70 cm.

LITERATURE CITED

Fry, F. E. J. 1951. A fractionating column to provide water of various dissolved oxygen content. Can. J. Technol. 29: 144–146.

Radford, E. P. 1964. The physics of gases. *In* Handbook of Physiology, W.O. Fenn & H. Rahn, eds. American Physiol. Soc., Washington. Section 3, Vol. I, pp. 125–152.

Rodeau, J. L., & A. Malan. 1981. An O_2–CO_2 mixing system for studies on water-breathing animals. J. Appl. Physiol. 51: 229–231.

Weast, R.C. 1970. Handbook of Chemistry and Physics. CRC Press, Cleveland, Ohio.

10

MEASUREMENT OF PHYSIOLOGICAL GASES IN BLOOD

10.1. INTRODUCTION

Although Priestley (ca. 1774) is generally credited with the discovery of oxygen, the first to appreciate its chemical properties and its critical role in animal respiration was undoubtedly Lavoisier. Between 1775 and his death by guillotine in 1794, Lavoisier and his colleagues conducted experiments demonstrating that respiration was indeed a form of combustion, involving the consumption of oxygen and concomitant release of carbon dioxide, and that the oxidative process was chemically the same as the combustion of charcoal and the oxidation of metals. Their mistaken belief that the combustion occurred in the lungs was disproved by Heinrich Gustav Magnus, when he demonstrated in 1837 that venous blood contained more CO_2 than arterial blood and that the reverse was true for oxygen. Magnus also developed the first practical, quantitative blood gas analyzer, which relied on vacuum extraction of gases, absorption of the CO_2 in caustic potash (KOH), and combustion of the oxygen with hydrogen. Since that time, a large number of "wet" chemical methods have been devised and used, as well as many methods taking advantage of other properties of the gases.

10.2. BLOOD AND PLASMA SAMPLING

10.2.1. Syringe Sampling

In order to obtain an accurate measurement of the quantity of any gas in blood, a sample must first be drawn in such a way as not to change the concentrations. While this seems a simple enough proposition, there are actually a number of errors that may be made, some obvious and others less so. The most common device for collection of blood samples is the syringe, which may be purchased in an almost infinite variety of sizes and shapes, the most common being plastic and glass. If samples of blood are to be stored in the collection syringe for any length of time, glass is probably the better choice, since the plastics used in syringe manufacture have an appreciable permeability to most gases, including O_2 and CO_2.

The principal problem with syringes is the *dead space*, that volume contained within the needle and syringe hub. In a microliter syringe this volume may be as little as 3 μl, and in a typical 1 ml tuberculin type syringe, about 0.03 to 0.04 ml, which in both cases is 3–4% of the syringe volume. If the syringe is to be drawn only half full, as much as 8% of the contained volume will be dead space. If the dead space is left full of air, that 8% can easily contain half as much oxygen as the sample being drawn, particularly if the sample is of venous blood or blood with a low oxygen capacity. Various materials have been employed to fill the dead space volume in syringes, mercury and mineral oil being two common ones. Mercury is less desirable due to its toxicity, the problems in handling it, its expense, and the fact that it does not wet the syringe walls. Mineral oil is far superior, but it is messy and may cause problems if it is subsequently expelled into various kinds of measurement apparatus. One alternative is to fill the dead space with a physiological saline, carefully removing all bubbles. Although the percentage volume error is still the same, and should be taken into account as dilution in any subsequent calculations, the gas capacity of the saline is an order of magnitude lower than that of an equal volume of air, so the likelihood of a significant change in the sample's gas concentration is less. In some cases it is possible to withdraw a partial sample into the syringe, mix the blood with the saline that was filling the dead space, expel the mixture, and then draw the sample desired after re-inserting the needle. This simple one-step dilution of the dead space fluid renders the dead space error almost negligible, but in some cases it is undesirable to waste the blood or to insert the needle twice. Yet another ploy is to draw the sample into the syringe with as little agitation and mixing as possible. When this is done, the saline from the dead space can usually be seen as a separate layer, and when aliquots are withdrawn from the syringe for further use, one can simply avoid mixing this saline into the aliquot withdrawn. With all of these different approaches, the most careful attention to details and the development of a smooth, consistent technique is of the greatest importance.

Another way in which sample errors may occur with syringe sampling of blood is due to the high shear which may occur in the needle and the high vacuum that may be exerted with a syringe plunger. Especially with small syringes (actually, those with a low ratio of barrel cross-section to needle cross-section), it is possible to produce such a high vacuum in the needle tip that turbulence and cavitation occur in the needle, resulting in partial degassing of the sample. This can usually, but not always, be observed as very fine bubbles appearing in the sample and will reduce the partial pressures of all gases in the sample. High flow velocities in the needle may also cause rupture of the red blood cells (if present) by excessive shear (Chapter 2). While this may not directly affect the total gas concentrations, it may radically change the partial pressure, since hemoglobin has a very different affinity in solution compared to that *in situ*.

The use of evacuated sample collection tubes (Vacutainers) has become increasingly common in clinical work, but these tubes are, of course, completely unsuitable for work with physiological gases.

10.2.2. Catheter Sampling

Chronically indwelling catheters are also used in many experimental situations for serial blood sampling, and many of the same problems may occur as with syringes. High vacuums on the attached sampling syringes are to be avoided, due to possible shear rupture and cavitation. The dead space in catheters is an even more serious problem, since the volume is much higher and the distance longer. Since the flow velocity profile in small tubes is so non-linear (Section 2.7.1 and Fig. 2.6), the center of the catheter lumen will be freed of saline long before the thin, low-velocity layer near the wall. A good procedure for ensuring that the saline is completely out of the catheter before sampling is to use a three-way sampling valve and a saline reservoir syringe, as shown in Fig. 10.1. About 1.5 times the catheter volume is first withdrawn into the smaller sampling syringe, the valve turned, and the sampling syringe discharged into the saline reservoir syringe. This process is then repeated one or more times until the entire catheter and the dead space in the valve and sampling syringe is filled with undiluted blood, whereupon the final sample may be drawn. The blood and saline discharged into the saline reservoir syringe are then returned to the animal, plus enough saline to re-fill the catheter. As the saline is withdrawn from the catheter in a

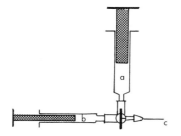

Fig. 10.1 Diagram of a three-way valve and syringe system for catheter sampling.

vertebrate, the blood becomes progressively redder, but the eye cannot be trusted to determine when the saline is 100% removed.

In some situations it may be desirable to sample very slowly from catheters, either with a syringe or by simply letting the blood flow out under its own pressure. At low flow rates with long catheters, the residence time of blood in the catheter may become long enough so that the permeability of the catheter wall is a significant factor. A typical PE60 catheter (0.030″ internal diameter × 0.048″ external diameter) has a wall thickness of only 0.009″, and significant gas exchange between the blood inside and the air may take place. (Techniques for implantation and maintenance of catheters are described in some of the references at the end of this chapter.)

10.2.3. Sample Transfer and Plasma Separation

When a large sample of blood needs to be split into sub-samples, the same opportunities for error exist as with the original sampling. Thus, some care in ensuring that the sample is not diluted, cells disrupted, or the sample exposed to air bubbles is required. One simple way to remove aliquots of a sample from a syringe is to remove the needle from the sampling syringe with a twisting motion, so as not to shake the tip and introduce bubbles, and then take sub-samples with a second syringe inserted into the hub opening of the sampling syringe. As blood is withdrawn from the sampling syringe, the plunger may be advanced at an equal rate, so that the surface of the original blood sample stays at the top of the hub. In this way the cross-sectional area of sample exposed to the air is held to a minimum, perhaps 1 mm², and the distance limitations of diffusion in aqueous media (Section 2.5.1) ensure that negligible error will result. Alternatively, a rubber serum stopper or similar closure device may be placed over the hub in place of the needle, and with a second syringe inserted through it and into the barrel of the sampling syringe, the sub-sampling syringe may be filled by gentle pressure on the sampling syringe's plunger. For syringes without removable needles, the original sampling syringe and the secondary sampling device may be connected by a short length of fine bore polyethylene tubing (of the sort used for catheters), but the dead space will probably have to be filled and rinsed once or twice to avoid dilution errors.

For the separation of plasma and cells, many techniques have been described. One useful method is to stopper the sampling syringe tightly and centrifuge the sample in the syringe. Usually some sort of clip or adaptor will be required to hold the plunger in place, and only low-speed centrifugation is possible, but the elimination of an extra transfer step is a decided advantage. More often the sample must be expelled from the sampling syringe into a centrifuge tube, so precautions against equilibration with air must be taken. The centrifuge tube may first be filled with a small quantity of mineral oil and the sample injected slowly beneath the oil layer. This precaution comes to naught, however, if bubbles are accidentally injected beneath the oil along with the blood, and the oil may interfere with subsequent handling and analysis. With long narrow tubes, it is often possible to omit the oil, provided that the tube is filled gently, with the needle below the surface of the blood sample; the surface area of the liquid in the top of the tube is small in relation to the volume; the centrifugation time is short; and the separated plasma or cells can be withdrawn from below the (air-exposed) surface layer. If only small amounts of separated plasma or cells are required, glass capillary tubes serve nicely, since the condition of small surface area is met, and the tubes can be broken to effect the separation following centrifugation.

10.3. OXYGEN IN BLOOD

10.3.1. Partial Pressure

The measurement of oxygen partial pressure (P_{O_2}) is performed almost exclusively with oxygen electrodes. In most common commercial instruments, the electrode is combined with a mechanical chamber that provides for minimum sample volume, a means of introducing the blood sample anaerobically, and temperature control, such as the system shown in Fig. 10.2. The design requirement for a small sample volume has three important consequences: the sample usually cannot be stirred, O_2 consumption by the electrode itself becomes important, and "capacitance" of the membrane electrolyte may become important. Electrodes for blood measurement, then, usually employ cathodes of very small surface area. A decade or so ago, these electrodes were very difficult to work with, since the electrical current generated is a function of the cathode area (see Chapter 4), and small-signal

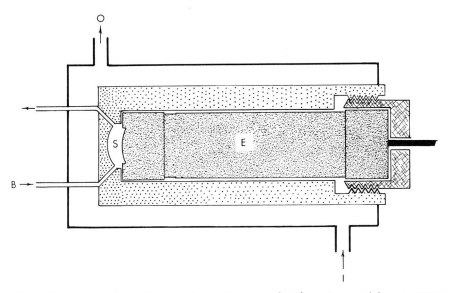

Fig. 10.2 Cross-sectional diagram of a temperature-jacketed cuvette containing an oxygen electrode used for blood measurement. The sample is introduced through (B) into a small volume sample chamber (S). The electrode (E) must have a small self-consumption rate to avoid significant errors.

DC amplifiers were difficult and expensive to design. The advent of new integrated circuits has made it practical to design electrodes with current output of 10^{-12} A/torr or less, and thus with an insignificantly low electrode O_2 consumption. The problem of electrode capacitance is not so obvious, but comes about because of the finite volume of electrolyte contained between the electrode membrane and the cathode surface. To take an extreme example, let us assume that the electrolyte and sample volumes are equal and that a fully oxygenated sample is injected right after a completely de-oxygenated one. At equilibrium, the sample partial pressure will be greatly reduced, and a "memory" artifact will be obtained. In practice, this problem is generally not very serious, since the electrolyte volume between the cathode and the membrane can be kept quite small, and the oxygen capacity of the sample is usually greater than that of the electrolyte. A precautionary test for this memory effect, however, is to simply inject a further quantity of the sample after the electrode has come to apparent equilibrium and observe whether there is any further change in the reading. If there is, and if the direction of the error is dependent on the value of the previous sample, then a memory or capacitance effect is present.

The electronics associated with the electrodes may occasionally become quite intricate, with microprocessors built in for automatic calibration, sample identification print-out, etc., but all that is required in principle is a polarizing voltage, a current amplifier, and some display circuitry, either digital or analog. The electronics will also provide for correction of the "zero current" and for adjustment of the full scale calibration for convenient units. Often there is also a quick test circuit for membrane leakage, which is a common problem with some batches of membrane material.

10.3.2. Indwelling Oxygen Microelectrodes

Increasing use has recently been made of indwelling microelectrodes for *in vivo* or *in situ* measurements of P_{O_2}. These microelectrodes are usually constructed as shown in Fig. 4.11, using an electrolytically etched platinum wire encased in a flexible polyethylene catheter. Since the electrodes are operated without a membrane, they are subject to "poisoning" with time, and tend to be somewhat flow sensitive. Each one must be calibrated frequently, and flow sensitivity should be checked. Since they also have a high output impedance, a preamplifier similar to that in Fig. 4.10 should be mounted as close as possible to the electrode.

A more complex needle microelectrode that incorporates both a reference and a membrane has been described by Baumgärtl and Lübbers (1983). Its performance specifications are impressive, but its construction and calibration represent a daunting task. The completed electrode has a tip diameter as small as 0.6 μm and has been used to measure micro-scale gradients of P_{O_2} in tissues such as the rat cerebral cortex.

10.3.3. Oxygen Content Measurement

The Tucker Method. A variant of the electrode-and-chamber methods described above for partial pressure measurements may also be used for O_2 content measurement (Tucker, 1967). The idea is to provide a larger volume sample chamber and to fill it prior to sample introduction with a solution that will convert chemically bound O_2 to physically dissolved O_2. The content may then be calculated from the chamber volume and the solubility coefficient for O_2 in the chamber fluid. Tucker's chamber apparatus is constructed like the CO_2 measurement chamber shown in Fig. 5.4 and may have a typical volume of 1 ml. The chamber is filled with a saponin-ferricyanide solution. The saponin (a detergent) serves to lyse the red blood cells, and the ferricyanide combines chemically with the heme group of hemoglobin, driving the oxygen into solution. The chamber is usually maintained at 30 to 40°C to speed analysis and to increase the sensitivity by reducing oxygen solubility (Chapter 2). Depending on the conditions and the chamber volume, the method may be used for samples in the range of 5 to 100 μl.

Manometric Methods.   These methods are little used nowadays, but were important in the early days of respiratory physiology and provide accurate results in practiced hands. The best-known survivor of these early methods is the Van Slyke apparatus, which was discussed in Chapter 4. The principle of the method is extraction of the oxygen after saponin-ferricyanide treatment, followed by measurement of the reduction of the collected gas volume upon absorption of the oxygen in pyrogallol.

Gas Chromatographic Methods.   From time to time other gas chromatographic devices have appeared on the market for measurement of blood oxygen, but these have never been very successful. In order for gas chromatography to yield good results, the bolus of gas to be analyzed should ideally be injected onto the beginning of the column in as small a volume as possible. When oxygen is extracted from blood by a gas stream flowing over it or bubbling through it, the extraction is slow, which tends to spread and skew the chromatographic peak that results. There is also a problem with water vapor, which interferes with the operation of some column packings. In general, it is an expensive and rather complicated way to measure blood oxygen.

Miscellaneous Methods.   Mass spectrometry has been employed for blood O_2 content measurement, more or less substituting the semi-permeable "leak" inlet for the oxygen electrode in the Tucker method. Due to the complex and expensive nature of the equipment involved, this method has not found wide application.

The measurement of percentage saturation of hemoglobin is sometimes confused with measurements of O_2 content, but the percentage saturation allows calculation of the content only when the concentration of hemoglobin or the total oxygen capacity of the sample is also known.

10.4. CARBON DIOXIDE IN BLOOD

10.4.1. Partial Pressure

Electrode Method.   The electrode-and-chamber method described above (Section 10.3.1 and Fig. 10.2) will also serve for CO_2 measurement when a CO_2 electrode is substituted for the O_2 electrode. There are some further problems in the measurement of P_{CO_2} in this fashion, however, that are more serious than in the measurement of O_2. The most serious problem is the slow reaction time of the electrode, which is a function of both the length of time required for partial pressure equilibration between sample and electrolyte and the reaction time of the bicarbonate buffer employed for the measurement inside the electrode (see Chapter 5). The primary use of the P_{CO_2} electrode is in mammalian studies, and at 37°C the equilibration time is about 1.5 min. The process has a Q_{10}, however, of about 2, so at 7°C the time required is around 12 min. The response time of the electrode is also strongly dependent on the membrane material used, with silicone membranes the fastest but sometimes the least stable.

Electrode capacitance, or "memory," is also more of a problem with CO_2 electrodes, since the electrolyte volume contained between the membrane and the pH-sensitive surface is generally larger than in an oxygen electrode, and the capacities of the sample and the electrolyte are not very different. The sample volume required is effectively increased by the need to inject the samples in two or three portions with equilibration time in between (Boutilier et al., 1978).

Finally, the P_{CO_2} electrode becomes quite unsatisfactory in much of the work with ectothermic and aquatic animals, since the measurement temperatures are low, as are the P_{CO_2} values to be measured, in the range of 0.5 to 5 torr. Since the P_{CO_2} electrode potential is a logarithmic function of the P_{CO_2} in the sample, the readings will also be off scale with most of the meters manufactured for mammalian or clinical applications. In these cases, the user is better off connecting the electrode to a standard pH meter and using a calibration graph of pH vs. P_{CO_2} to calculate the actual sample P_{CO_2}. It is also rather difficult to obtain good calibration points under these conditions, since often the electrode drift will occur with nearly the same time constant as the measurements (i.e., 12 to 15 min). With the most careful attention to drift correction, calibration, and memory effects, the P_{CO_2} electrode can provide reasonable measurements at low temperatures and P_{CO_2} values, but the precision will be considerably less than that advertised for mammalian conditions.

Equilibration or "Astrup" Methods. Since pH is a much easier measurement to make than P_{CO_2}, and since the relationship between pH and P_{CO_2} is linear on a semi-log scale over reasonable ranges of P_{CO_2} (Fig. 5.3), a method for estimation of P_{CO_2} was developed using a two-point equilibration. The idea is to simply split a sample in half and to equilibrate one half to a P_{CO_2} above that expected in the sample and the other half to a P_{CO_2} below that expected in the sample. After equilibration, the pH of the two portions is measured and plotted as in Fig. 5.3. If the pH of the original sample is also known, then the P_{CO_2} may be estimated directly from the graph. With the proper apparatus, the equilibration can be performed with small samples in a fairly short time, so the method is not as tedious as it sounds. The disadvantages are that evaporation of the sample may induce errors and the errors are compounded logarithmically, due to the nature of the relationship in Fig. 5.3.

Indirect Calculation Methods.   A third way to measure P_{CO_2} in blood is not to measure it at all, but rather to calculate it from the Henderson–Hasselbalch equation for the carbonic acid system (see Chapter 5). In order to perform the calculation, one must know the total CO_2 content, the pH, and the "pseudo-constant," pK', at the appropriate pH, temperature, and ionic strength. For human blood these values are well known, and a wide variety of nomograms, slide rule calculators, computer programs, etc. are available for easy calculation of P_{CO_2} from total CO_2 content and pH. For other animals, especially when the conditions are low temperature, low P_{CO_2}, high pH, and variable ionic strength, the value of pK' is quite uncertain. The errors in this method are also logarithmically cumulative, so small errors in pH or pK' estimation may lead to quite large errors in the calculated P_{CO_2}. It is best to verify the value of pK' under the conditions of the experiment.

Was the Blood Sample in Equilibrium?   Normally, the slow equilibration reaction between CO_2 and HCO_3^- is speeded up by several orders of magnitude in (vertebrate) blood due to the presence of the enzyme *carbonic anhydrase* in the red blood cells. In most invertebrates, however, there is no carbonic anhydrase activity in the circulating blood, so the possibility should not be overlooked that the blood *in vivo* was not in equilibrium. By removing a sample and allowing it to come to equilibrium in a sampling electrode chamber, the blood is also allowed to recover from any disequilibrium that may have been present. Theoretical studies, not so far verified experimentally, indicate that this may be the case for most blood samples from crabs, for example, so that most of the literature values do not accurately represent the *in vivo* P_{CO_2} values (Cameron, 1978). Where this is a problem, special kinetic techniques must be developed in the future.

10.4.2. Carbon Dioxide Content

Manometric Methods.   The remarks about manometric analysis of O_2 also apply to the measurement of total (i.e., dissolved plus chemically combined) CO_2 in blood. That is, these methods are little used now but were important historically, and may still provide excellent results in practiced hands with a minimum equipment investment. The analysis of CO_2 is based on the same vacuum extraction in acidified blood, followed by measurement of the reduction of the collected gas volume upon absorption of the CO_2 in a strong base, usually KOH. The acidification step is common to most methods of CO_2 analysis and makes use of the equilibria shown for the carbonic acid system in Chapter 5. Strong acidification tends to drive these reactions to the left, favoring the formation of CO_2 gas. At a pH of 4 or below, virtually all the CO_2 in solution is in the dissolved gas form, and none is combined as carbonate or bicarbonate.

The Cuvette Method. A system for total CO_2 analysis analogous to the Tucker method used for O_2 (above) has been described (Cameron, 1971) utilizing a CO_2 electrode in place of the O_2 electrode and an acid solution in place of the saponin-ferricyanide solution (Fig. 5.4). This method also works better at slightly elevated temperatures and is applicable to μl sample volumes. Actually, it seems quite feasible to combine the two methods, using a chamber with two opposed electrodes, one for O_2 and the other for CO_2, and an acidic saponin-ferricyanide solution. So far no one seems to have tried it, but it should provide both CO_2 and O_2 content analyses on the same small sample.

Conductometric Methods. A combination of acidification to facilitate extraction of CO_2 from the liquid to the gas phase and absorption of the CO_2 from the gas phase back to an alkaline liquid phase may be used in conjunction with conductometric detection for CO_2 analysis (Cameron, 1981). The advantages of this method are that the conductometric detector can be extremely sensitive to small quantities of CO_2 and samples can be processed in a relatively short time. The general principles of conductometric detection were discussed in Chapter 5.

Gas Chromatographic Methods. Gas chromatography is somewhat more successful for CO_2 than for O_2, since the CO_2 can be collected in a cold trap for injection onto the column in a small volume. The most practical scheme is to extract the total CO_2 from the blood sample by acidification in a carrier gas stream, collect the CO_2 generated by freezing it in a liquid N_2 trap, and then shunt the collected CO_2 onto the column using a six-port sampling valve, as shown in Fig. 5.6. This method can also be extremely sensitive and is appropriate for micro samples, but requires fairly elaborate apparatus and a ready supply of liquid nitrogen.

Miscellaneous Methods. Virtually any of the methods for measurement of CO_2 in gases described in Chapter 5 may be used with blood, provided some appropriate "plumbing" is constructed to first acidify the sample and strip the dissolved CO_2 into a carrier gas. The amounts of CO_2 actually present, however, are usually quite small, so the sensitivity of the method may be a consideration. A 20 μl sample of typical fish blood, for example, which may contain 8 mM L^{-1} total CO_2, has a total content of only 0.16 μM, or 3.6 μl. If this is added to a gas stream flowing at 50 cc min^{-1} over a 2 min period, the concentration of CO_2 in the gas stream is around 36 ppm, and whatever detection method is employed must be able to measure concentrations in that range with reasonable reproducibility and accuracy.

10.5. OTHER GASES

10.5.1. Carbon Monoxide

The vacuum extraction plus chemical absorption methods of Van Slyke and Neill (1924) may be employed for the measurement of carbon monoxide (CO), as well as O_2 and CO_2. The extraction procedure is the same, except that sometimes a more acidic ferricyanide solution is used to ensure cleavage of the fairly tight bond between hemoglobin and CO. After absorption of the O_2 and CO_2, the volume of gas remaining consists of N_2 plus CO. The CO is then absorbed with several portions of $CuCl_2$, with which CO forms a loose bond. The CO present in the sample is measured manometrically as the difference in extracted gas volume before and after absorption. Several other methods are now available for the measurement of CO, all of which first require that the CO be scrubbed from the liquid into a gas phase either by vacuum or carrier gas extraction. Gas chromatography is quite suitable (see Chapter 4), as is mass spectrometry. The combination of CO with hemoglobin is quite stable and has a higher affinity bond than Hb–O_2; the degree of saturation of Hb with CO may be assessed spectrophotometrically.

10.5.2. Hydrogen

In the mid-19th century, hydrogen was determined most directly by combustion with oxygen, since the molar ratio is exactly 2:1. This reaction is so exergonic (heat-yielding) that a high explosion hazard exists, and special care to make the gas measurements before and after combustion isothermally is necessary. Hence, the method is little used today, and one of several physical measurements of H_2 are more commonly used. Gas chromatography is an excellent choice, since H_2 has diffusivity, thermal conductivity, and a molecular size very different from that of the other common atmospheric or physiological gases. Mass spectrometry is also a sensitive and accurate

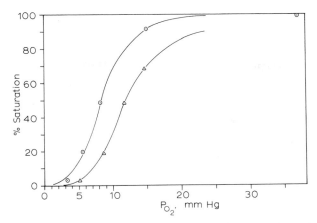

Fig. 10.3 A typical oxygen dissociation curve for fish blood at 10 and 20°C. (Cameron, unpublished data.)

method for H_2 detection. For direct measurement of H_2 in solution without extraction into a gas phase, the oxygen electrode (Chapter 4) may be used by reversing the polarizing voltage. The cathodic reaction is then

$$H_2 = 2 H^+ + 2e^-$$ (Eq. 10.1)

and the electrode will operate under conditions generally similar to those discussed for O_2 electrodes. This technique is sometimes used with microelectrodes when a single breath of a gas containing H_2 is used as an indicator in circulation studies.

10.6. DISSOCIATION CURVE MEASUREMENT

10.6.1. Oxygen Dissociation Curves

The typical blood oxygen dissociation curve, shown in Fig. 10.3, is a graphic representation of the non-linear relationship between the partial pressure of O_2 and the O_2 content. For historical reasons, the O_2 content is often expressed as "Vol %," which is the content in milliliters of gas STPD contained in 100 ml of blood. The abcissae may also be expressed in percentage saturation, but in this case the total capacity (i.e., the content at atmospheric O_2 partial pressure) must also be known in order to calculate the content at any given partial pressure. Since the P_{CO_2} affects the dissociation curve for oxygen, this must also be measured and/or controlled.

Measurement of the O_2 dissociation curve may be approached in two ways: by controlling the partial pressure and measuring the content, or by

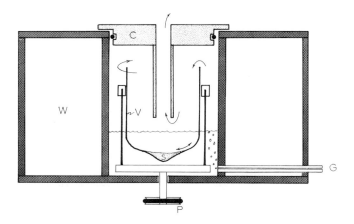

Fig. 10.4 Diagram of a rotating blood equilibrator. The equilibrating gas (G) must bubble through isothermal water (W) and flow over the rotating blood film (S) before escaping through a removable cover (C). The equilibration vessel (V) alternately rotates and stops for more effective mixing, and the entire apparatus is encased in a temperature jacket.

controlling the content and measuring the partial pressure. The earliest method employed was to equilibrate the blood at a series of partial pressures and then measure the content at each with the Van Slyke apparatus. In the following discussion, equilibration techniques and problems are described, followed by a description of the so-called mixing technique and the dynamic methods used in commercial dissociation curve analyzers (DCAs).

Equilibration.   Since the diffusion of gases through bulk solutions is very slow (Chapter 2), equilibrating a blood sample by simply allowing gas to flow over the top of it would take an unacceptably long time. Many different types of apparatus have been devised to speed the equilibration of gases with blood, all of which have the aim of exposing the blood in a relatively thin layer by rotation or shaking. One of the most effective types is shown in Fig. 10.4, and is either available commercially (from Instrumentation Laboratories) or can easily be constructed. This device spreads the blood into a thin film on the wall of the central glass vessel by intermittent rapid spinning. In equilibrating blood, many problems may occur, the most obvious of which are evaporation or dilution. The gas flowing through the equilibration device must be saturated with water vapor at the same temperature as the sample. If it is not saturated or is saturated at a higher temperature, evaporation of the sample will result. On the other hand, if the gas is saturated at a lower temperature, moisture will condense as the gas is warmed in the equilibrator, and dilution of the blood sample will result. For the most critical work, the gas should be equilibrated with a solution of approximately the same ionic strength as the blood, since ionic strength influences vapor pressure (Chapter 2).

Equilibration for long periods may also lead to metabolic or chemical changes in the blood. Red blood cells may accumulate acidic end products, intracellular phosphate concentrations may change, or metabolic CO_2 may accumulate, all of which change the oxygen affinity. Invertebrate bloods may be subject to polymerization or dissociation of the pigment, denaturation of the proteins, or formation of other aggregates with lipid and protein. Equilibration times should therefore be kept as short as possible, provided that the completeness of equilibration can be verified. Fairly long times are sometimes unavoidable; a pigment of very high O_2 affinity may be difficult to completely de-oxygenate, for example, since a partial pressure gradient of only a few torr will maintain considerable saturation of the pigment. In vertebrate bloods, hemolysis may occur with long equilibration times; this should be watched for by periodically centrifuging a small aliquot in a micro-hematocrit tube and visually checking the plasma for pigment.

The Mixing Technique. A very quick and convenient technique for measuring oxygen dissociation curves was described by Edwards and Martin (1966). The procedure is as follows: a blood sample is split into two roughly equal portions, one of which is fully oxygenated in an equilibration device, the other completely de-oxygenated. When equilibrium has been attained in both, portions of each are drawn one after the other into a single syringe in a carefully measured ratio. A drop of mercury is also drawn into the syringe, enough to seal the end and provide a bead for mixing the sample by rocking the syringe back and forth. The mixed sample is then injected into an oxygen electrode such as the one shown in Fig. 10.2, and the O_2 partial pressure is measured. The percentage saturation of the sample is determined by the ratio of the volumes of oxygenated and de-oxygenated blood drawn into the syringe. The beauty of this method is that partial pressure is generally much easier to measure than content, and once the two aliquots are equilibrated, the entire curve may be constructed in a fairly short time. By equilibrating the two halves with gases with various partial pressures of CO_2 added, the effect of the Bohr shift may also be assessed. A final measurement of the total oxygen capacity is needed, however, in order to express the dissociation curve in terms of content, rather than percentage saturation. The dead space volume of the syringe must also be taken carefully into account when withdrawing the blood into the syringe. To get a 1:1 ratio in a 1 ml syringe with 0.03 ml of dead space, for example, the first portion should be drawn up to 0.47 ml (after filling the dead space with the sample and getting rid of bubbles)

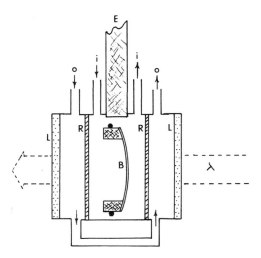

Fig. 10.5 Cross-sectional diagram of Reeves' (1980) dissociation curve analyzer cell. The blood sample is held in a thin film (B) between two thin Teflon membranes and equilibrated with an O_2-free gas circulated through an inner chamber (i). The outer chamber (o) contains air or oxygen separated by two oxygen-permeable membranes (R). An oxygen electrode (E) monitors the P_{O_2} in the inner chamber, and the spectrophotometer's light beam passes through the lucite end windows (L) and the blood film. (Redrawn with permission from Respir. Physiol.)

and the second portion to 0.97 ml in order to have 0.50 ml of each. If the first portion were withdrawn up to the 0.50 mark, the ratio would be 1.06 to 1.00 and an appreciable error in the dissociation curve would result.

Dynamic Methods.   Several dynamic methods and apparati have been described for the measurement of dissociation curves, but the one that appears to offer the most advantages over other methods is that described by Reeves (1980, 1984). The principles of the method are more easily understood by referring to the diagram of the apparatus employed (Fig. 10.5). Blood is spread in a thin film between two 6 μm Teflon membranes, and this "sandwich" is placed in the center of a chamber small enough to fit in the sample compartment of a dual wavelength spectrophotometer. The sandwich sits in a center compartment that also has an oxygen electrode inserted into it to monitor the partial pressure of O_2. The center compartment is separated from two outer compartments by gas-permeable silicone membranes, and both compartments are connected to gas sources with inflow and outflow tubes. Initially the inner and outer compartments are flushed with nitrogen (zero O_2) plus the desired CO_2 concentration, and the ratio of hemoglobin absorbance at 430 and 453 nm is measured. The outer compartments are then flushed rapidly with oxygen plus the same CO_2 concentration, whereupon the oxygen begins to diffuse into the inner compartment. The oxygen electrode in the inner compartment records the rise in partial pressure, which provides the ordinate for the dissociation curve, and the changing absorbance ratio of 430 to 453, which is a function of hemoglobin percentage saturation, provides the abcissa. The complete curve may be determined in about 4 min on a very small volume of blood. The only disadvantage of this method appears to be the requirement of the rather intricate chamber and the dual-wavelength spectrophotometer.

Somewhat similar methods, based on simultaneous measurement of O_2 partial pressure in the blood and addition of O_2 to the sample by diffusion through a membrane, have led to two commercial dissociation curve analyzers, the Aminco "Hemoscan" and the Radiometer-Copenhagen DCA-1. Both of these require much larger blood samples, and consequently rather long pre-equilibration times with nitrogen–CO_2 mixtures, which produce some of the problems discussed in connection with equilibration (above).

The choice of method for the measurement of O_2 dissociation curves, then, depends on the equipment and sample volume available and on the number of times these measurements need to be made. If all that is required is a basic set of dissociation curves for other kinds of studies, then the mixing technique is probably the best choice. On the other hand, if many measurements need to be made under a variety of conditions, the expense or time required to develop one of the dynamic methods is probably justified.

10.6.2. Buffer Curves and CO_2 Dissociation Curves

The requirements for the measurement of CO_2 dissociation curves (or *combining* curves) and buffer curves are similar to those for the measurement of O_2 dissociation curves. Blood must be equilibrated with gases containing a range of CO_2 partial pressures, and then either the total CO_2 content or the pH is determined, generating graphic relationships like those shown in Fig. 10.6. The main problems that can occur are similar to those that arise in studying O_2 dissociation, namely, proper humidity of the equilibration gas and the effects of long equilibration on the blood properties. The oxygenation state should be carefully controlled, since with most blood pigments the buffer capacity is a function of O_2 binding to the pigment. Careful temperature control is also required. At present, no equivalent of either the mixing methods or the dynamic methods described for O_2 is available, so the determination of these CO_2-related blood properties is simply a matter of equilibrating and then determining the content or pH by any of the methods described for those measurements (Chapters 5 and 6).

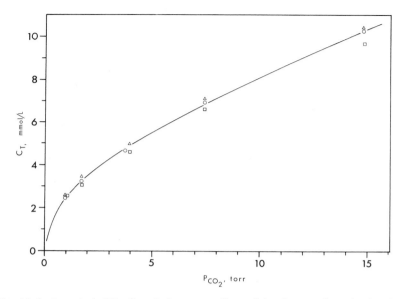

Fig. 10.6   A typical CO_2 dissociation curve (for a fish; three replicates), showing the relationship between P_{CO_2} and total CO_2 content (CT) in mmol L^{-1} at $15°C$.

LITERATURE CITED

Baumgärtl, H., & D. W. Lübbers. 1983. Microcoaxial needle sensor for polarographic measurement of local O_2 pressure in the cellular range of living tissues. Its construction and properties. *In* Polarographic Oxygen Sensors. E. Gnaiger & H. Forstner, eds. Springer-Verlag, New York. pp. 37–65.

Boutilier, R. G., D. J. Randall, & G. Shelton. 1978. Some response characteristics of CO_2 electrodes. Respir. Physiol. 32: 381–388.

Cameron, J. N. 1971. A rapid method for the determination of total carbon dioxide in small blood samples. J. Appl. Physiol. 31: 632–634.

Cameron, J. N. 1978. Excretion of CO_2 in water-breathing animals: a short review. Mar. Biol. Letters 1: 1–12.

Cameron, J. N. 1981. U.S. Patent No. 4,321,545. Carbon dioxide measurement system.

Edwards, M. J., & R. J. Martin. 1966. Mixing technique for the oxygen–hemoglobin equilibrium and Bohr effect. J. Appl. Physiol. 21: 1898–1902.

Grubb, B. R., & C. D. Mills. 1981. Blood oxygen content in microliter samples using an easy-to-build galvanic oxygen cell. J. Appl. Physiol. 50: 456–464.

Hersch, P. A. 1964. Galvanic analysis. Adv. Anal. Chem. & Instr. 3: 183–249.

Reeves, R. B. 1980. A rapid micro method for obtaining oxygen equilibrium curves on whole blood. Respir. Physiol. 42: 299–316.

Reeves, R. B. 1984. Oxygen equilibrium curves of whole blood determined by micro dynamic thin-film technique. *In* Techniques in the Life Sciences, Part 4, Vol. II, Respiration Physiology.

Tucker, V. A. 1967. Method for oxygen content and dissociation curves on microliter blood samples. J. Appl. Physiol. 23: 410–414.
Van Slyke, D. D., & J. M. Neill. 1924. The determination of gases in blood and other solutions by vacuum extraction and manometric measurement. I. J. Biol. Chem. 61: 523–573.

SUGGESTED FURTHER READING

Hoar, W. S., & D. J. Randall, eds. Fish Physiology. Academic Press, New York. (Various volumes from 1970 through 1984, including methods sections in appendix to Volume 10A, 1984).
Popovic, V., & P. Popovic. 1960. Permanent cannulation of aorta and vena cava of rats and ground squirrels. J. Appl. Physiol. 15: 727–728.
Soivio, A., K. Nyholm, & K. Westman. 1975. A technique for repeated sampling of the blood of individual resting fish. J. Exp. Biol. 63: 207–218.

CHAPTER

11

FLOW AND VELOCITY MEASUREMENT

11.1. FLOW MEASUREMENT IN GASES

11.1.1. Thermal Flowmeters: Hot-Wire and Thermistor

The use of a hot-wire element as an arm of a resistance bridge was discussed in Sections 3.5 and 4.3.3 as a method of detecting different gases based upon their thermal conductivity. In that case, the flow was maintained constant and the variable measured was the changing thermal conductivity of the gases coming from the separation column. If the composition of the gas is constant but the flow rate varies, the hot-wire sensor may also be employed as a flow-measuring device. One possible circuit arrangement is given in Fig. 11.1. In this circuit, the hot-wire sensor element is exposed to the gas flow and is heated by a solid-state regulator (7808) configured as a constant current source. As the wire is cooled by increasing flow, its resistance, and hence the voltage across it, changes and is amplified and processed by the two operational amplifiers.

An alternate hot-wire circuit is shown in Fig. 11.2, where one sensor element is exposed to the gas flow and the other is employed as a reference element in a bridge circuit to balance out changes in ambient temperature. The bridge current is controlled by the pass transistor so that the hot-wire element operates isothermally, and the bridge current is a function of the gas flow over the sensor element. The advantages of this type of flow sensor

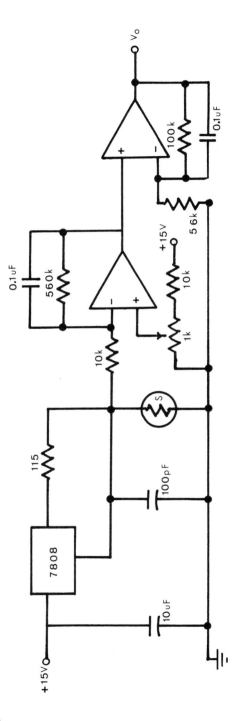

Fig. 11.1 A circuit for employing a hot-wire sensor element (S) in a constant-current configuration for measurement of flow. The op amps may be any common type, such as the LM741CN, and the 7808 is a solid state voltage regulator of the LM7808CT series. Other component values as shown. The output at V_0 is a non-linear function of flow.

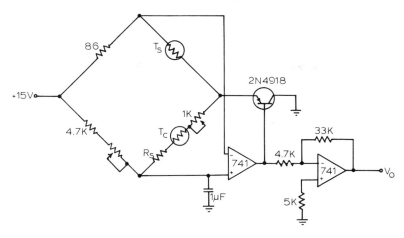

Fig. 11.2 A Wheatstone bridge circuit employing two hot-wire sensors, one (T_s) as the flow sensor, the other (T_c) as an ambient temperature compensator. The output of this circuit is also non-linear. The values of R_s and the 1K trimmer are adjusted for zero output at zero flow. (Utenick, 1971.)

are that it is relatively cheap and easy to construct and has a very fast response time. Using a 0.025 mm 90% platinum–10% iridium wire, it is possible to measure flow variations of up to 2 kHz, which is far more than adequate for physiological studies. A major disadvantage is that the output signal is non-linear. It is difficult to construct analog electronics to linearize the output, since the flow/output function is quite complex, but by interfacing the output to a laboratory micro-computer, it should be relatively easy to program a linear conversion based on polynomial fitting of an empirical calibration curve.

Another feature of the hot-wire flowmeter is that there is no directional sensitivity. For some applications this is a serious drawback, for example in respiratory studies, where inspiratory and expiratory flow must be distinguished. A clever variation of the hot-wire flowmeter has been described that has directional sensitivity (Yoshiya *et al.*, 1975), but both the construction of the sensor and the associated electronics are much more complex than the simple types shown in Figs. 11.1 and 11.2.

By selecting the proper values of reference voltage and resistance across the bridge arms, the hot-wire sensors may be replaced by small thermistor beads. A thermistor is a device with a predictable and fairly linear negative temperature coefficient, i.e., as temperature rises, its resistance falls. There are really no particular advantages to using a thermistor instead of a resistance wire, except perhaps the lower power requirement of the former. On the negative side, thermistors are more expensive, somewhat more difficult to

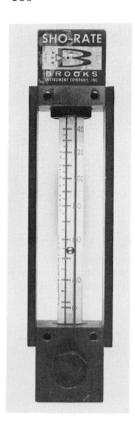

Fig. 11.3 A ball-type flowmeter. With modifications of bore diameter and float material, these meters can handle a wide range of both gas and liquid flows.

replace, and have a much slower time constant, since their mass and thermal inertia are much greater than those of a fine wire.

11.1.2. Mechanical Gas Flowmeters

The commonest gas flowmeters are those based on a simple ball or float in a tapered tube, such as the one shown in Fig. 11.3. These are generally inexpensive, but not very precise, so they are suitable for applications in which a high degree of accuracy is not required. More accurate models may be obtained, but the extra precision required in construction increases the cost substantially. A flowmeter of this type is only usable over a fairly narrow range of flow, so for measurements over a wide range, one must either have several of them or have a type which has interchangeable barrels and floats

Fig 11.4 Schematic diagram of the principle of a pneumotach. Gas flowing through a resistance (R) will cause a pressure drop between P_1 and P_2 proportional to the flow.

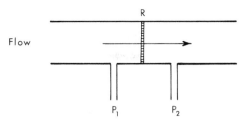

for accommodating different flow ranges. The calibration of these meters is given for a particular gas, usually air, at a specified temperature, usually 20°C. To use them under any other conditions requires that a correction be applied, since the viscosity of the gas depends upon its composition and temperature. They are also subject to errors from contamination; anything which changes the flow resistance of the barrel wall, the float, or any of the orifices in the flowmeter will shift the calibration. For critical uses, a fine particulate filter ahead of the meter will reduce these problems.

Another type of mechanical flowmeter may be constructed with paddle or propeller mechanisms, usually contained within a surrounding flow channel. The cumulative flow through such a device may be recorded by a revolution counter connected to the propeller, or may be translated to a flow per unit time basis by either mechanical or electrical means. These meters have the advantage of being usable over fairly wide ranges of flow and are easy to use as summing (integrating) meters. On the other hand, they are somewhat fragile, may be quite expensive, and are subject to the same variations with gas temperature and composition as the float type.

11.1.3. The Pneumotach

A simple pneumotach is shown in schematic form in Fig. 11.4. It consists of a flow tube, a source of a small resistance to flow (in this case a screen), and two pressure measurement ports which are connected to a differential pressure transducer (see Chapter 12). The principle is based upon the Poiseuille–Hagen equation for pressure (P) and flow (F):

$$F = \Delta P \pi R^4 / 8 \eta l \qquad \text{(Eq. 11.1)}$$

where η is viscosity, l length, and R radius. Since everything in this equation is constant for the device in Fig. 11.4 except the pressure gradient (ΔP) and the flow, one should be a linear function of the other. Pneumotachs can be constructed for nearly any flow range, flow channel shape and size, and may

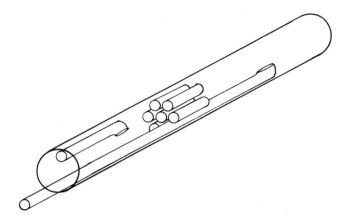

Fig. 11.5 Diagram of a simple disposable pneumotach that can be made from polyethylene catheter tubing. (Courtesy of R. G. Boutilier.)

be constructed from disposable materials, making them very inexpensive to use. For example, short pieces of small bore plastic tubing may be packed in the center of a larger and longer piece of plastic tubing, and two small tubes glued in perpendicular to the large tube on either side of the section packed with small tubes. This makes a slight flow resistance, due to the restriction of the small bore tubing sections, which can be sensed through the two perpendicular ports. The whole device constitutes a throw-away pneumotach which could be used for small animal studies (Fig. 11.5).

11.1.4. The J-Valve

One of the oldest devices, and still a useful one for measuring air flow in physiology is the so-called J-valve, shown in Fig. 11.6. It is a T-tube with leaflet valves arranged in such a way that air flowing in one direction (e.g., inspiration) enters through one arm, and air flowing in the other direction exits through the other arm. By connecting the expired arm to an analyzer, large air bag, etc., the expired air may be measured or analyzed. The advantages of this system are that the streams of gas in two directions may be kept separate, except for the dead space volume in the valve and connecting tubes; the flow resistance of the valve system may be kept quite low; and by scaling the size of the valve to the system (animal) in which it is used, virtually any flow range may be accommodated.

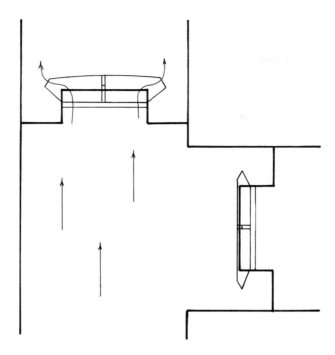

Fig. 11.6 Cross-sectional diagram of a J-valve. This type of valve offers very low flow resistance, and in a three-way configuration like the one shown, it can be used for separating inspired and expired gases.

11.2. MEASUREMENT OF FLOW IN LIQUIDS

11.2.1. Thermal Methods

Virtually the same hot-wire and thermistor devices described in Section 11.1.1 may also be used for the measurement of liquid flow, given the modifications necessary to compensate for the different mechanical properties of the media. That is, hot-wire sensors must be sufficiently robust to stand the mechanical forces of fluid flow, and the current dissipation characteristics of the bridge circuits used must be adjusted for the high heat capacity of water and other liquids compared to gases. Very simple probes may be constructed on this basis, and have been used for sensing flow velocity or for measuring animal activity (Brumbley & Arbus, 1979).

A note on velocity and flow measurement is in order. Strictly speaking, most of the sensors discussed above actually respond to the velocity of flow immediately adjacent to the sensor surface. In a tube of fixed geometry, there will also be a fixed relationship between the velocity at a given point in the stream (i.e., where the sensor is placed) and the total *flow*, which is usually taken to mean a volume per unit time. The empirical calibration process usually employed with these sensors may obscure this relationship, but when using thermistors or hot wires as probes, it must be kept in mind that only velocity is sensed and that converting the velocity to the flow requires other information as well.

11.2.2. Electromagnetic Flowmeters

Motion of an electrically conducting material through a magnetic field in a direction other than along the lines of magnetic force will induce an EMF (voltage) in the moving material. This principle has been exploited in the design of electromagnetic flowmeters for physiological use. The sensor design of a blood flowmeter based on this principle is shown in Fig. 11.7a. It consists of an electromagnet shaped to surround a flow channel and two electrodes contacting the lumen of the transducer. When the flowmeter is placed around a blood vessel, the flow of (electrically conductive) blood through the lumen sets up a voltage which is measured by the electrodes. The activation of the electromagnet is usually pulsatile, either in sine-wave or square-wave form, and extra field windings and electrodes may be added by some manufacturers to assist in zero flow calibration.

The signal from the electromagnetic sensor is a function of the field strength and the electrical conductivity of the blood. The latter does not vary much, but calibration of each probe is usually carried out with the section of vessel *post mortem*. Application of these probes, shown in Fig. 11.7b, is not always easy, as obtaining a reliable zero flow calibration point can be troublesome, and a good fit between the vessel wall and the probe must be obtained. If the probe is too small, flow will be constricted, and if it is too large, the vessel will only contact the electrodes during peak pressure, and artifacts will occur. Practically, one must have a large number of probes of different sizes to fit a particular blood vessel, and since the probes are each quite expensive, the total measurement system may cost $10,000 or more.

The best results with electromagnetic flow probes are obtained from chronic implants; enough time is allowed between implantation and measurement so that the probe becomes encapsulated in connective or scar tissue. This ensures relatively little motion of the probe, thus reducing artifacts in the recordings due to pressure changes in the vessels. Such chronic implants, with encapsulation of the probe, are very difficult to obtain in ectothermic animals, and the main usefulness of the method has so far been in mammalian and

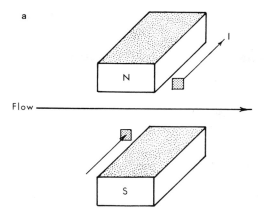

Flow ⟶

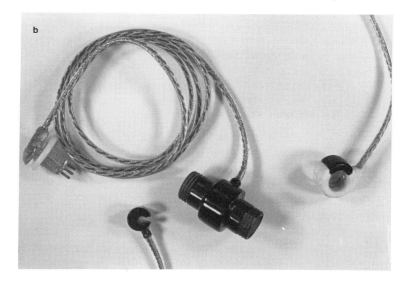

Fig. 11.7 (a) The principle of electromagnetic flow measurement is illustrated. Flow of an inductive material (such as blood) through a magnetic field will generate an electrical current (*I*) proportional to the flow rate. (b) Some commercially available electromagnetic flow probes. The platinum pick-up may be seen on the inside of the probe on the right. The one in the center is a cannulating type.

bird studies. Application to cardiac output measurement in fish, for example, has been tried by several investigators (various personal communications), but without much success. In addition to the problem of poor healing of implants, the ventral aorta, which is the site required for total cardiac output measurement in fishes, is very short and changes dimension so much during the cardiac cycle that it makes electromagnetic probes almost impossible to use.

The electromagnetic flow probe may also be used to measure sea water flow and perhaps in some relatively salty fresh water. One such application has been to measure the ventilation volumes of crabs (Johansen *et al.*, 1970) and fish (Piiper *et al.*, 1977). For these applications, a large lumen diameter probe is used so as not to add significant flow resistance. This may occasionally be a problem, since most probes of 10 to 20 mm inside diameter are manufactured with mammalian applications in mind. A 20 mm probe might be used on a cow aorta, e.g., where the flow is very high, so the manufacturer will use a low magnet inductance in order not to generate too high a voltage at the pickup electrodes. For measurement of seawater flow rates from a crab, however, what is required is a large lumen diameter and a high inductance, so that low flow rates may be measured accurately. For such specialized applications, it may occasionally be necessary to have a probe custom made or to make one's own probes.

11.2.3. Ultrasonic Doppler Effect Velocimeters

The example used in introductory physics textbooks to illustrate the Doppler effect is the rise and fall in the tone (frequency) of a train's whistle as the train travels first toward and then away from the listener. The amount of frequency change is a function of the train's velocity relative to the listener. In the schematic of a Doppler effect velocity sensor given in Fig. 11.8, we see that there is a sound source and a crystal sound receiver (comparable to the listener). The sound waves, which are usually in the inaudible ultrasonic range, are not transmitted in a straight line to the receiver, but are bounced back to the receiver from particles in the flow stream between them. In practice, the sending crystal is pulsed at high frequency, and the frequency of the sounds "heard" by the receiving crystal are compared with those transmitted. The frequency shift is a function of the particle velocity with respect to transmitter and receiver.

In physiological applications, these Doppler sensors are usually molded into a cuff-like device that fits over a blood vessel and is sutured in place.

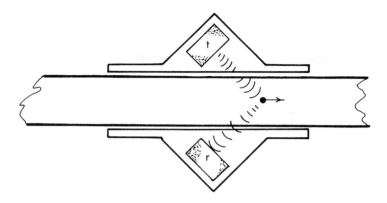

Fig. 11.8 Schematic diagram of a Doppler flow probe. The transmitting crystal (t) and the receiving crystal (r) may be combined in some types. Sound waves bouncing off particles in the flow stream will be frequency-shifted by an amount proportional to their velocity with respect to the crystals.

The particles reflecting the sound waves are the blood cells moving in the vessel lumen, so the sensor will give a directional signal; that is, the frequency will shift up when the blood flow is toward the receiver and down when away. In practice, the received signal will not be a pure (frequency-shifted) tone but a spectrum of frequencies, reflecting the non-uniform velocity profile of blood flow in vessels (see Section 2.7.1), so the electrical circuitry must average the frequency of the received signal. In order to translate this velocity-dependent signal into flow, the sensor must be calibrated *post mortem* using the actual section of blood vessel over a range of known flows.

One advantage of the Doppler flowmeter over the electromagnetic type is that it does not depend on electrical contact between the vessel wall and pickup electrodes, which is a source of artifacts with electromagnetic probes. There is less need to fit the transducer exactly to the vessel, then, and this cuts down greatly on the cost of many different-sized probes. The probe must not allow too much shift of vessel position within the sensor lumen, however, since this will affect the velocity/flow calibration.

11.2.4. Laser Doppler Velocimeters

Recent advances in fiber optics and laser technology have led to the development of an interesting new method for flow measurement. Laser light striking moving particles is scattered and reflected back with a frequency shift

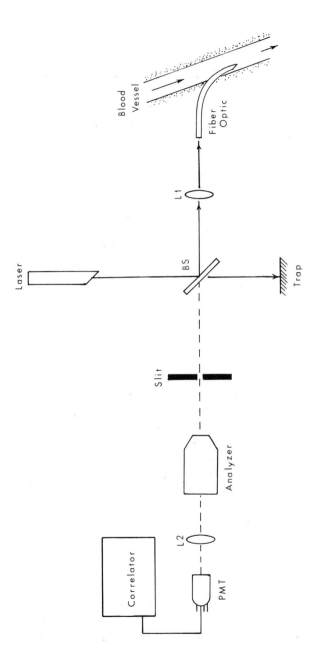

Fig. 11.9 Functional block diagram of a laser Doppler velocimeter system for measuring blood flow. The operation of this apparatus is described in detail by Tanaka (1980). The measurement depends on frequency shifts of the laser light reflected at the tip of the fiber optic catheter. (Reprinted with permission from Tanaka, 1980. Copyright CRC Press, Inc., Boca Raton, Florida.)

that is proportional to the velocity (v) of the particles:

$$\Delta f = f - f_\mu \qquad\qquad \text{(Eq. 11.2)}$$

and:

$$\Delta f = 2v \cos(a/w) \qquad\qquad \text{(Eq. 11.3)}$$

where a is the scattering angle and w the wavelength. For physiological flow velocities, the change in frequency is very small, only about 3 parts in 10^{10}, but by mixing the scattered light with the original beam, a heterodyne "beat" can be produced and measured with precision. An arrangement for measuring blood flow with a fiber optic probe is shown in Fig. 11.9, taken from Tanaka (1980). Using this apparatus, Tanaka has been able to measure velocity and flow in a rabbit vein; venous flow rates are generally below the effective range of other measurement techniques. This technique has also been applied to a range of other flow measurement problems and may become commercially available in the next few years.

LITERATURE CITED

Brumbley, D. R., & E. A. Arbus. 1979. An inexpensive hot-wire anemometer suitable for behavioral research. Comp. Biochem. Physiol. 69A: 449–450.

Johansen, K., C. Lenfant, & T. A. Mecklenburg. 1970. Respiration in the crab, *Cancer magister*. Zeits. f. vergl. Physiol. 70: 1–19.

LaBarbera, M., & S. Vogel. 1976. An inexpensive thermistor flowmeter for aquatic biology. Limnol. Oceanogr. 21: 750–756.

Piiper, J., M. Meyer, H. Worth, & H. Willmer. 1977. Respiration and circulation during swimming activity in the dogfish *Scyliorhinus stellaris*. Respir. Physiol. 30: 221–239.

Tanaka, T. 1980. Laser Doppler velocimetry using fiber optics with applications to transparent flow and cardiovascular circulation. *In* Physical Sensors for Biomedical Applications. M. R. Neuman *et al.*, eds. CRC Press, Boca Raton, Fla. pp. 119–133.

Utenick, M. R. 1971. Design of a hot-wire anemometer. Instr. Soc. Amer. Trans. 10: 21–28.

Yoshiya, I., T. Nakajima, I. Nagai, & S. Jitsukawa. 1975. A bidirectional respiratory flowmeter using the hot-wire principle. J. Appl. Physiol. 38: 360–365.

CHAPTER 12.

FORCE, DISPLACEMENT, AND PRESSURE MEASUREMENT

12.1. BASIC UNITS

Force. In the cgs system, the unit of force is the *dyne*, which is defined as the force necessary to accelerate a mass of 1 g by 1 cm sec^{-2}. In the SI system, the unit is the *Newton*, defined as the force necessary to accelerate a 1 kg mass by 1 m sec^{-2}. Since 100 cm = 1 m and 1000 g = 1 kg, 10^5 dynes = 1 Newton (symbol N).

Pressure. Pressure is a force applied uniformly over an area, and there are a number of different units in common use. The SI unit is the *Pascal* (symbol Pa), defined as 1 Newton per cm^2, i.e., 1 Pa = 1 N cm^2. The Pascal is quite inconveniently small, a standard atmosphere equaling 101,325 Pa, so the most common form encountered is the kiloPascal (kPa). In the cgs system, the basic pressure unit is the *barye*, defined as 1 dyne cm^2, but the *bar*, which equals 10^6 baryes, is much more commonly used, as are its divisions, the millibar and microbar.

There are also several non-sanctioned units of pressure in common use. In many kinds of work, including physiology, it is often convenient to express pressures in terms of the height of a column of either mercury or water that

179

the pressure will support. Thus, the mm Hg is a common pressure unit, since so many pieces of apparatus employ mercury columns or barometers for pressure measurement. Atmospheric pressures are commonly reported in mm Hg; $\frac{1}{760}$th of a standard atmosphere is defined as 1 *torr*, which is the same as 1 mm Hg. To convert from one type of unit to the other, the equation is 1 torr = 133.322 Pa at 0°C and standard gravity. For precise measurements, corrections for temperature (as it affects mercury density) and local gravity (largely a function of latitude) must be made, and most mercury barometers are supplied with correction tables. As an example, the usual correction of the mercury barometer in the author's laboratory in Texas is about − 4.1 torr. For very small pressures and some physiological apparatus, it is more convenient to express pressure in terms of the height of a column of water or saline. These units may be converted using the equation 1 mm Hg = 13.6 mm H_2O = 13.1 mm saline.

12.2. DISPLACEMENT TRANSDUCERS

Large-scale displacements may be rather easily measured by a variety of either mechanical or electrical devices. One early physiological device was the smoked-drum kymograph, in which a mechanical arm drew a line in the smoked covering of a drum as the drum was turned at a known rate. These primitive but functional devices are still in use in some teaching labs, where they have the virtues of low cost and a clearly viewed and understood mechanism. Two simple electrical devices may be made with either round or linear potentiometers, provided, if necessary, with springs to aid in return after displacement. By applying a voltage to either end of the potentiometer, the voltage of the wiper arm becomes a linear function of the displacement of the mechanical shaft.

In some applications it is important to measure displacement without adding mechanical resistance to the system, and for these, spring devices are clearly inappropriate. Light may be used in a variety of ways, including simple occlusion of a beam, or with mirrors arranged in appropriate ways.

For very small displacements, none of the methods discussed above may be appropriate, but various others are available. The principle of one such method is that the capacitance of a capacitor constructed with two conducting elements separated by a gap of any dielectric material will be a function of the distance separating the two elements. If such a capacitor system is employed as the resonating capacitor in a tuned oscillator, the frequency of oscillation will vary as the separation between the two conducting elements varies. With proper selection of materials and tuning of the circuit, very small

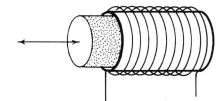

Fig. 12.1 Use of a ferrite slug tuning coil for linear displacement. Movement of the slug into or out of the coil winding field changes the inductance of the coil and can be converted to a linear electrical signal.

changes in the separation distance (i.e., displacement) may be measured with high accuracy. A variant of this principle has been applied to the measurement of changes in blood vessel diameters during the cardiac cycle, for example.

Inductance as well as capacitance may also be used for the measurement of displacement, based on the same principle as tuning coils in many kinds of radio gear. The illustration in Fig. 12.1 shows a ferrite "slug" used to vary the inductance of a coil. Very small movements of the slug into or out of the field of the coil will induce frequency shifts in an oscillator incorporating the coil winding, and these frequency shifts can be measured with an accuracy of 1 ppm or less, providing extremely sensitive displacement measurement. The system is one-dimensional, however, which limits its application.

Another possibility for measurement of displacement is the use of either radar or sonar. In either, the idea is to measure the time delay between propagation of a pulse, either a radio wave or a sound wave, and the return echo. If the speed of travel in the medium is known, the exact distance between transmitter and target may be calculated. The sonar ranging system used in Polaroid cameras is available to experimenters and could be applied to displacement measurements in a variety of situations (Ciarcia, 1985).

Two- and three-dimensional displacements are much more difficult to measure, but fortunately are needed only rarely. One way to avoid the rather cumbersome mechanical arrangements necessary for measurement along two perpendicular lines of displacement is to substitute measurement of distance and angle from a fixed sensor, as shown in Fig. 12.2. This system can easily be extended to a three-dimensional problem by measuring a second angle, the azimuth angle. Calculation of displacement in each of three perpendicular directions is then a straightforward problem in trigonometry. The most theoretically elegant solution is to use an accelerometer system which will register acceleration in the X, Y, and Z directions. The mathematical derivation of position at any instant in time from the three acceleration vectors requires some calculus. Acceleration is the second derivative of position, so a first integration provides velocity vectors. A second integration procedure provides the position function. With digital computer techniques, this is relatively simple to do in real time. Great advances in miniaturization of accelerometers for such applications as sensing heart motion have been made

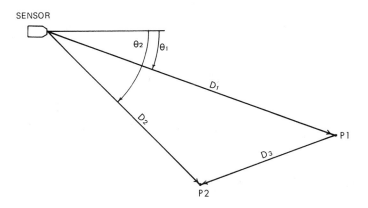

Fig. 12.2 Distance plus angle method of measuring two-dimensional displacement. The displacement distance (D3) can be calculated using simple trigonometry from the two measured distances (D1 and D2) and the difference in angles.

in recent years. One type of micro-accelerometer is shown in Fig. 12.3 (Angell, 1980).

12.3. PRESSURE TRANSDUCERS

A typical physiological pressure transducer is shown in Fig. 12.4. In operation, the chamber over the diaphragm is filled with fluid and connected by tubing to the point where pressure is to be measured. Since water is for all practical purposes incompressible, pressure applied at the end of the connecting tube will be faithfully transmitted to the sensing diaphragm of the transducer. There are a variety of ways to translate pressure on the diaphragm to an electrical signal, but again the most common method is to connect resistive or inductive elements of a bridge circuit (see Chapter 3) to the diaphragm. One type employs quartz fibers as arms of a resistive Wheatstone bridge circuit; changes in tension on the quartz fibers cause changes in the electrical resistance of the fibers. Most bridge transducers have two or four elements of the bridge connected in reciprocal pairs in order to reduce temperature instability. That is, changes in temperature affect both elements of a pair equally, canceling out the effect. From simple hydraulic principles, we see that in use the transducer diaphragm must be at the level of the tip of the connecting tube; otherwise a true zero cannot be obtained.

Transducers of the type shown are quite reliable for static pressure measurements, but for faithful registry of dynamic pressure waves, several precautions must be observed. The presence of any gas bubbles in the chamber or connecting tubes will drastically reduce the frequency response of the

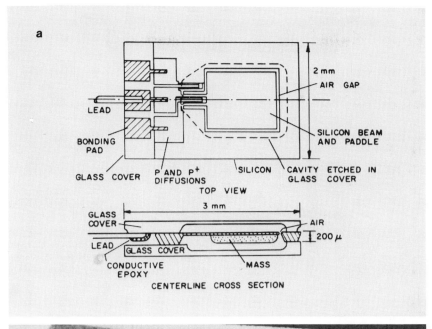

Fig. 12.3 Schematic diagram (a) and photograph (b) of a micromachined accelerometer. An accelerating force acting on the suspended mass causes bending of the silicon paddle, which can be converted to a linear electrical signal. (Reprinted with permission from Angell, 1980. Copyright CRC Press, Inc., Boca Raton, Florida.)

Fig. 12.4 A diaphragm strain gauge pressure tranducer (Statham). The transparent dome is filled with fluid and connected by fluid-filled tubes to the site where pressure measurements are to be obtained. Beneath the clear dome there is a metal diaphragm whose displacement is translated into changes in the resistance of arms of a Wheatstone bridge.

system, causing a slower rise in the signal and a reduced peak pressure. This is called *damping* (see Fig. 12.5). Elasticity of the walls of the connecting tubing will cause a similar effect, so the tubing should be as short and rigid as possible. In many cases it is desirable to make most of the connecting length out of fine-bore metal tubing, using flexible material only for the terminal piece. In addition to the elasticity introduced by bubbles or by tubing walls, there is an additional effect of the volume displacement allowed by diaphragm movement. When pressure is changed in a step at the catheter tip, the increased pressure will cause an inward deflection of the diaphragm, increasing the volume of the system. There will then be bulk fluid flow through the connecting tubing, and the resistance (viscous drag) of the tubing will cause damping. Since the resistance is inversely proportional to the fourth power of the radius, this effect can become very serious with small bore catheter tubing (such as PE50). The length of the tubing also affects the

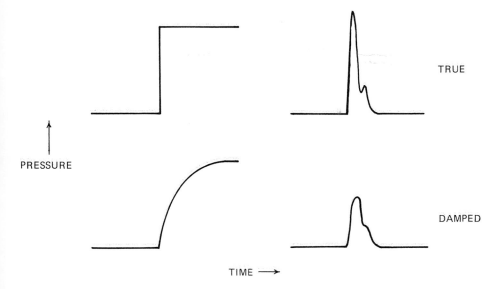

PRESSURE

TIME ⟶

TRUE

DAMPED

Fig. 12.5 Illustration of the effects of damping on pressure wave recording. The true pressure is shown at the top, a square wave change at the left, and a typical arterial trace at the right. The traces below show the effects of damping.

resistance, so keeping the tubing as short and as large bored as possible is required for faithful pressure recording.

A final problem in pressure recording is movement and flow artifacts. A pressure catheter placed in an artery facing the direction of flow will experience a higher pressure than one facing downstream. Likewise, the Venturi effect may reduce the recorded pressure in a catheter whose tip is perpendicular to the flow. Assessment of the seriousness of this placement error should be carried out in the particular system of interest. Movement of the connecting tubing introduces artifact signals from both compression of the tubing walls and inertia of the liquid contained in them. For some applications, this means that implantable transducers of a type not sensitive to movement artifacts are the only acceptable method.

12.3.1. Implantable Transducers

Technological developments of the last 10 years have led to the introduction of several new pressure transducer types, some of which have great advantages over the type described above. One of these is based upon fiber optic technology and is shown in cross-section in Fig. 12.6. This transducer consists

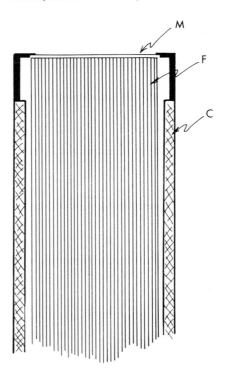

Fig. 12.6 A fiber optic pressure transducer. Light transmitted along fiber optics (F) is reflected back from a thin glass membrane (M). As pressure deforms the membrane, less light is reflected directly back into the fibers, thus yielding a pressure signal. The fiber optics are enclosed in an opaque catheter (C).

of a bundle of optical fibers, about half of which are used to transmit light and the other half to receive it. The end of the fiber optic bundle is cut off flat, and an air gap of 50 to 100 μm is left between the cut ends and a thin glass membrane that is sealed over the tip of the catheter sheath. Changes in pressure on the outer surface of the membrane cause changes in the amount of light reflected back through the receiving fibers, and this change has been found to be linear over a physiologically useful range.

Another type of transducer depends upon changes in electrical conductivity in semiconductor layers. Using "micromachining" methods developed in the electronics industry, transducers with a variety of configurations may be constructed in implantable form. These newer types are largely replacing piezo-electric devices, which depend upon changes in the resonant frequency of quartz crystals under varying pressure. Most of these recent developments have not yet been translated into commercial products, but may be on the market in the next few years.

LITERATURE CITED

Angell, J. B. 1980. Transducers for in vivo measurement of force, strain and motion. *In* Physical Sensors for Biomedical Applications, M. R. Neuman *et al.*, eds. CRC Press, Boca Raton, Fla. pp. 45–54. (See also other articles in this volume.)

Ciarcia, S. 1985. Living in a sensible environment. A collection of alarm and monitoring circuits. Byte 10(7): 141–158.

MEASUREMENT OF IONS AND SOLUTION PROPERTIES

13.1. OSMOTIC PRESSURE

The osmotic pressure of a solution is determined by the number of active particles dissolved per unit volume, as discussed in Section 2.8.1, and is usually measured indirectly by the change in one of the related colligative properties of water (freezing point depression, boiling point increase, or vapor pressure depression). The oldest and still the most common method depends upon measurement of the freezing point of the solution, which can be done with various instruments. One simple and inexpensive way is to simply cool a sample of the solution in a bath and observe the temperature at which ice crystals form. This is a tedious method. Great care must be taken to avoid supercooling of the solution, which requires a very slow cooling rate. It is easy to generate small temperature gradients between the sample tube and the bath, and small temperature errors lead to rather large errors in estimation of the osmolarity, since the freezing point is depressed by only 1.86°C for each mole of active dissolved particles.

One common type of osmometer takes advantage of the freezing point depression and another property of water, the latent heat of crystallization. As water is cooled toward 0°C, it takes approximately 1 calorie per gram per °C to effect the temperature change. The transition from liquid to solid

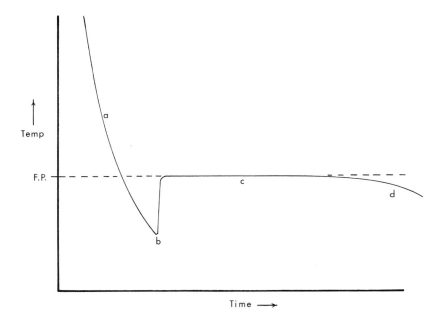

Fig. 13.1   Time–temperature graph for the Advanced Instruments freezing point osmometer. The sample is initially warm and is cooled rapidly along curve *a*. The rapid cooling induces supercooling, and at *b* freezing is initiated. The temperature rises to a plateau value at *c* which is maintained as long as the freezing process continues. Eventually the freezing is complete and the temperature falls again at *d*. The plateau value (*c*) is the true freezing point.

at 0°C, however, requires an additional 76.4 cal/g of heat loss. In one instrument, manufactured by Advanced Instruments, the sample is placed in a tube immersed in a very cold bath, so that a fairly high rate of heat loss occurs. When cooled rapidly in this way, the sample becomes supercooled, at which point a vibrating wire triggers freezing in the supercooled sample. During the period in which ice crystals are forming, a steady state temperature is reached at which the rate of heat loss to the bath is just equal to the rate of heat release from the crystallization. The time–temperature trace appears like that of Fig. 13.1, and the temperature measured at point C is a true measure of the osmolarity of the solution. A linear calibration curve is obtained with known-osmolarity standards. This instrument provides an osmolarity estimate of high accuracy in only a few minutes but has the disadvantage of requiring a minimum sample volume of 0.25 ml.

More recently, an instrument has become available that uses a very similar procedure based upon the decrease in vapor pressure. In the Wescor vapor pressure osmometer, a 5–8μl sample is sealed in a chamber containing a

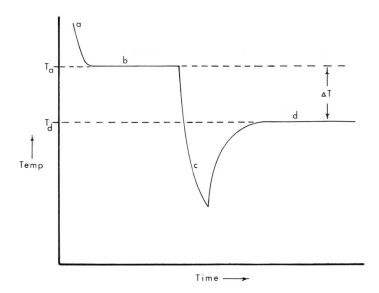

Fig. 13.2 Time–temperature graph for the Wescor vapor pressure osmometer. The initially warm sample is first equilibrated to the ambient chamber temperature (b). A cooling current drops the sensor below the dewpoint at (c), after which the sensor is allowed to gradually warm. The heat of vaporization of moisture formed on the sensor holds the sensor temperature at the dewpoint (d), and the difference between ambient temperature (b) and the dewpoint (d) is directly related to the osmolarity of the sample.

thermocouple. Passing an electrical current through the thermocouple cools it below the dewpoint of the chamber, which has first been allowed to come to equilibrium with the liquid sample. The cooling is then stopped, and the thermocouple temperature rises until a steady state is reached between heat gained from the ambient chamber and heat lost by evaporation of the moisture film on the thermocouple. The time trace of thermocouple temperature is shown in Fig. 13.2, which highlights the similarity between this method and the freezing point depression method described above. Since the heat of vaporization of water is 540 cal/g, the steady state temperature is reached even more easily during the evaporation of the water from the thermocouple and lasts longer, allowing a very accurate determination of the change in vapor pressure.

The vapor pressure osmometer allows a significant reduction of sample volume and is generally a rapid and convenient instrument. The thermocouple employed for the measurement, however, is quite delicate and must be kept clean by special procedures for best performance.

13.2. SALINITY MEASUREMENT

13.2.1. Salinity and Chlorinity

Salinity is defined as the total amount of solids dissolved in a water sample after all the carbonates have been converted to oxides, all bromide and iodide have been replaced by chloride, and all the organic matter has been oxidized. Salinity is therefore an artificial concept and is smaller than the total filtrable residue. Although sea water is predominantly an NaCl solution, there are numerous minor components (Table 2.7) that make the chemical measurement of total salinity a complex task. It has long been appreciated, however, that the relative proportions of the different salts are constant for waters of the open oceans, so measurement of any particular ion should allow the estimation of any of the others or of the total. Chlorinity is defined empirically as follows:

$$\text{Salinity} = 0.03 + 1.805(\text{chlorinity}) \qquad \text{(Eq. 13.1)}$$

where both are in g kg^{-1} H_2O or parts per thousand.

13.2.2. Titration

The commonest method for chemical measurement is the titration of chloride with silver nitrate (APHA, 1976). The silver nitrate forms an insoluble precipitate with Cl^-, and the measurement may be made either by observing the volume of silver nitrate taken to reach an end point with an indicator or by weighing the AgCl recovered by filtration and drying.

13.2.3. Electrical Conductivity

Since the ions dissolved in sea water render it electrically conductive in proportion to their total quantity and charge, salinity may be assessed by measurement of conductivity or resistance. In practice, however, this may not be a very satisfactory method, since the relationship between conductivity and salinity is non-linear and is strongly dependent upon temperature. As an approximate method, it may suffice, since it is quick and easy. The use of a series of known standards at the same temperature as the sample increases the accuracy considerably (e.g., Pollak, 1954).

13.2.4. Specific Gravity

A very inexpensive apparatus for measurement of salinity is the hydrometer (Fig. 13.3), which measures the specific gravity of the sample. For sea water at 15°C, a density of 1.026 corresponds to a salinity of 35 ppt. In practice,

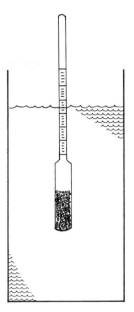

Fig. 13.3 A diagram of a hydrometer. The lower bulb is filled with some heavy material, and the upper portion is calibrated so that the depth at which it floats is a function of the liquid density.

the density read from the hydrometer is corrected to a true density from temperature correction tables and then converted to salinity with a second set of tables (APHA, 1976). In order to cover a wide range of salinity, it is often necessary to have a set of two or three hydrometers, since each one generally covers only a narrow range. Careful use of this method will provide estimates of salinity to within ±0.5 ppt.

13.2.5. Refractive Index

The refractive index of water, i.e., the amount by which light waves are bent or retarded when passing through it, depends upon the concentration of solutes added to it. For pure sodium chloride, the refractive index varies from 1.3330 to 1.3391 between 0 and 35 ppt, and this difference can be measured with fair accuracy with an optical device. The commonest such instrument is the American Optical Company's hand-held refractometer (Fig. 13.4), which is offered with absolute refractive index scales or specialized scales, including one for salinity. The precision of this method is not great, perhaps ±1 ppt, but it is adequate for many purposes, and of course is extremely quick and convenient for field use.

The refractometer is applicable to measurement of total dissolved solids

Fig. 13.4 A hand-held refractometer manufactured by American Optical. The sample is placed between the plastic cover and a special prism, and the refraction of the sample is measured on an optical scale through the eyepiece at the right. Refractometers are available with several scales, including one for salinity.

in other solutions as well, such as urine. Since protein has a much greater effect on refractive index than inorganic salts, total protein may also be measured reasonably well with the refractometer, provided the variations in inorganic salts are not too large. This provides a quick and easy method for assessing total serum or plasma protein, for example, with only a single drop.

13.2.6. Inductance

For high precision measurements of salinity, particularly in oceanographic applications, the method of choice is the inductance salinometer (Brown & Hamon, 1961). The principle of operation is that the inductance of an electrolyte solution is a function of the electrolyte concentration. In the Beckman instrument, the sample to be measured is used to fill a chamber so that it forms the core of a transformer winding, and thus influences the inductance of the transformer (Fig. 13.5). The inductor is used as one element of a bridge circuit, which contains both inductive and resistive elements, and is energized by a high frequency AC voltage, rather than a DC voltage such as is used in the simpler Wheatstone bridge (Section 3.5). The instrument is calibrated with standard seawater samples supplied by an oceanographic organization in Denmark, and tables are used to convert sample readings to true salinity. With care, results accurate to four or more decimal places may be obtained, which is quite important in oceanographic work but rarely of any significance in physiology.

13.3. MEASUREMENT OF WATER CONTENT

The obvious method of choice for measurement of water content of whole animals, tissues, etc. is to simply weigh them wet, dry them in some way, and weigh the dried residue. The difference gives the initial water content.

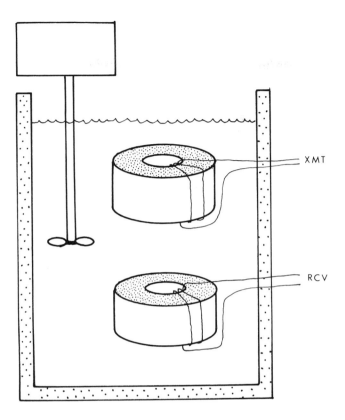

Fig. 13.5 Diagram of the sample chamber of an inductance salinometer. The stirred sample surrounds two coils, a transmitter (XMT) and a receiver coil (RCV). The signal induced in the receiver is a function of the inductance of the sample, which can be accurately related to the salinity.

Some care must be taken in the drying to ensure that other materials are not also lost, as for example might occur with volatile lipids dried at a high temperature. Many biological materials are also quite hygroscopic when dried, so care must also be taken to weigh the dried material before any significant absorption of moisture from the atmosphere can occur.

Drying in this fashion, of course, is a terminal method that allows only one measurement, and in some cases it may be important to make a series of measurements over time in the same tissue or animal. For these applications, tracer or indicator methods have been used, either with heavy water (D_2O—deuterium oxide) or tritiated water (THO—$^3H^1HO$). A full discussion of indicator or tracer methods is deferred to Chapter 14. The D_2O method is seldom used, since the tritium method is much simpler, and the analysis of deuterium with any precision is a tricky business.

13.4. MEASUREMENT OF SPECIFIC INORGANIC IONS

13.4.1. Sodium

Flame Photometry. In elementary chemistry probably everyone has dipped a platinum or nichrome wire into a solution of NaCl and observed the intense orange color when the wire was subsequently heated in a flame. Each element has a characteristic spectrum of light emitted in response to thermal excitation, and measurement of the intensity of that emitted light forms the basis for quantitative analysis by flame photometry. In commercial instruments, the components are a pressure and flow regulating system for the burner gases, an aspirator or atomizer for dispersing the sample in the flame, an optical system for isolating the desired wavelength of light, a detector for collecting and measuring the light, and electronics for amplification, scaling, and display (Fig. 13.6). In most instruments, the burner head is built in such a way as to incorporate the aspirating and atomizing functions, ensuring even dispersal of the sample in the flame. Optical systems vary considerably, the simplest being a filter allowing only a narrow spectral band to pass. More sophisticated systems may include various combinations of prisms, monochromators, or interference filters.

One inherent problem in flame photometry is that the intensity of the spectral line, for example from sodium, depends on the concentration of sodium in the sample and on the rate of aspiration of the sample. Therefore small variations in the aspiration rate, which might be caused by particulate contamination of the sample or by slight pressure fluctuations, must be compensated for. The usual way to do this is to add a known concentration of another element to the sample with an intense emission band that does not overlap the unknown. Then the relative emission in the two bands will yield the concentration of the unknown by a simple ratio. In practice, lithium salts are usually used to "swamp" the sample, and the ratio of the purple lithium emission to the orange sodium line is used to estimate the sodium concentration. Most modern machines also integrate or average the ratio over some time period sufficient to eliminate noise and random heterogeneities in the sample. Flame photometry is generally the method of choice for sodium analysis; it is quick, accurate, requires only 10–20 μl of sample, accommodates a wide range of liquids, and is relatively free of interference. With the addition of automatic sample changer accessories and printer output, the method is highly suitable for automated analysis of large numbers of samples.

Atomic Absorption. In the flame (emission) photometer described above, only about 1% of the atoms of any given element are in the excited (emitting) state at any time; the rest remain in the gas environment of the flame in a

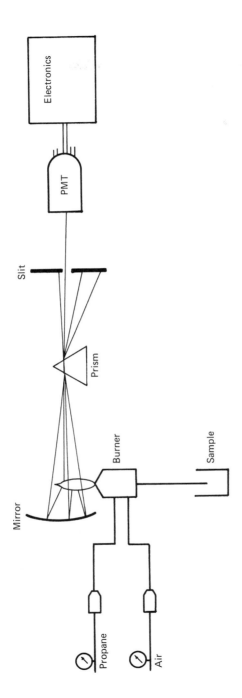

Fig. 13.6 Functional block diagram of the principal components of a flame photometer. See text for discussion of the operation.

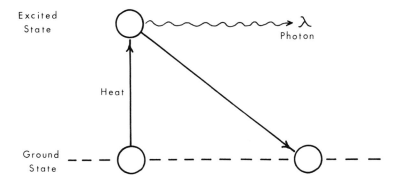

Fig. 13.7 Schematic diagram of the excitation principle of flame (emission) photometry. Atoms excited in the flame emit light of a characteristic wavelength when they subsequently decay to the ground state. Similarly, the wavelengths absorbed in the excitation process also have a characteristic spectrum, which may be utilized in an atomic absorption spectrophotometer.

well-dispersed state. These unexcited atoms may be excited by capture of photons having wavelengths characteristic of the element. Thus the atomic absorption principle is complementary to the flame emission principle; the latter depends upon a sharp emission line when excited atoms decay to the ground state (Fig. 13.7), the former upon a narrow absorption line when ground state atoms are excited.

The design of an atomic absorption instrument is similar to that of a flame photometer, except that a light source must be added. In order to reduce the problem of having to measure changes in a narrow spectral band against a continuous background spectrum, it is highly advantageous to use a nearly monochromatic source. The usual approach is to employ a hollow cathode lamp in which the cathode well is lined with the element to be measured, yielding a nearly monochromatic light at the desired wavelength. The concentration of an element in a solution is a function of the concentration in the sample being aspirated, the rate of aspiration, and the path length of the light beam through the flame. To increase sensitivity, the path length is usually elongated by the use of slotted burner heads. Variations in aspiration rate, however, are more difficult to handle, since the swamping technique employed in flame photometry is of no utility. Greater care in ensuring a constant aspiration rate is necessary for atomic absorption analysis. An additional source of error is variations in the lamp intensity, which are usually compensated for by splitting the beam, either by a mechanical beam chopper or a split mirror. One side is used as a reference, the other as a sample beam, and the electronics are designed to measure the ratio between the two beam halves' intensities.

Atomic absorption spectrophotometry is the preferred method for sodium only for samples of low concentration. For flame photometry, samples are usually diluted about 1 to 200 for a range of concentrations between 10 and 200 mmol L^{-1}, but for atomic absorption the dilution would have to be 10 to 100 times greater, since the sensitivity is much higher. Sodium is a ubiquitous contaminant, so at these extreme dilutions it is difficult to obtain results not influenced by sodium contamination of the diluent. There is also the necessity to either dilute in two steps, which introduces volumetric errors, or dilute to a very large volume, which is often inconvenient.

Other Sodium Analysis Methods. Sodium can be analyzed gravimetrically by precipitation as sodium zinc uranyl acetate hexahydrate (APHA, 1976), but this method is seldom used now. Sodium electrodes have become available and are good for some applications (see below).

13.4.2. Potassium

Potassium analysis in physiological fluids is nearly always conducted along with sodium analysis by flame photometry. The commercial instruments are often designed for (human) clinical use and may provide less accuracy in the determination of potassium, since it is often present at concentrations only about 3–5% as great as those of sodium. For greater accuracy, it may be necessary to perform a separate dilution of the sample. For the same reasons given for sodium analysis, the analysis of potassium by atomic absorption is generally not preferred, unless the concentrations to be measured are very low and the required dilution is not as great. Other methods for potassium measurement include a colorimetric test (APHA, 1976) and specific ion electrodes (see below).

13.4.3. Chloride

One method for measurement of chloride ion, silver nitrate titration, has already been discussed in connection with salinity measurement (Section 13.2.2), and a variant of this technique forms the basis for many commercial instruments for measurement of the chloride concentration in physiological fluids. The earliest of these was the Buchler–Cotlove chloridometer, which uses a titration vessel shown in Fig. 13.8. The sample to be analyzed is diluted in a solution of nitric and acetic acids, placed in a small titration vessel, and then raised by a sample holder so that four electrodes and a stirrer are immersed in the solution. When the titration is started, two of the electrodes

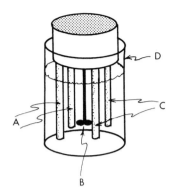

Fig. 13.8 Diagram of the titration vessel (D) of a Buchler chloridometer, showing the sensor (C) and generator electrodes (A). The sample is stirred by a small paddle (B).

are used to pass a precisely controlled constant current, which generates silver ions in the solution by the reaction

$$Ag \rightarrow Ag^+ + e^-$$

(Eq. 13.2)

As long as free Cl^- ions remain in solution, the Ag^+ ions generated combine with them and precipitate as AgCl, but when the Cl^- is exhausted, the conductivity of the solution starts to rise (Fig. 13.9). By measuring the time taken for the conductivity to rise by a certain amount and comparing this time for standards and unknowns, the chloride concentration can be calculated. This technique is known as *amperometric* titration. Various newer instruments employ variants of this principle to measure chloride, and may include microprocessor control and digital display.

13.4.4. Calcium and Magnesium

For both of these ions, atomic absorption spectrophotometry is the method of choice, and often a combination lamp may be used for both elements, though the dilution required may not be the same. To minimize interferences of various sorts, calcium particularly is often analyzed in a solution containing both HCl and a lanthanum salt, either $LaCl_3$ or La_2O_3, at a concentration of about 1% w/v.

For some applications, especially for field measurements, a titration procedure incorporating ethylenediaminetetraacetic acid (EDTA) as a chelator may be used. The EDTA combines first with calcium and then with magnesium in solution, but if the pH is made sufficiently high (about 12 to 13) to precipitate the magnesium as $Mg(OH)_2$, the EDTA may be used with any of several indicators to titrate only calcium. If the titration is also performed at a lower pH value, the difference will give an estimation of the magnesium concentration (APHA, 1976).

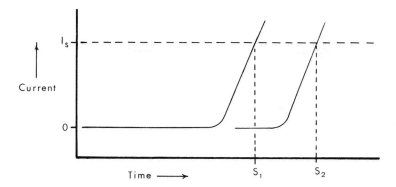

Fig. 13.9 The time–current graph for the Buchler amperometric chloride titrator. A cut-off current (I_s) is set to stop the instrument's timer, and the length of time required to reach the cut-off current for various samples (e.g., S_1 and S_2) is a function of the total chloride ion content.

13.4.5. Phosphate

No fully suitable instrumental or electrode methods have been developed for inorganic phosphate analysis, so the basic method described by Fiske and Subbarow (1925) is still in routine use. It requires that a series of reactions be carried out, resulting first in the formation of phosphomolybdate and finally in a phosphomolybdenum blue complex, which is measured spectrophotometrically. Several variations of this method are suitable for physiological fluids (Sigma Chemical Co., 1981; APHA, 1976), but most require removal of interfering organic substances first. For blood serum or plasma, for example, lipid and protein fractions must first be removed by treatment with trichloroacetic acid and centrifugation. For routine processing of large numbers of samples, these chemical procedures may be performed on automated analyzers from various manufacturers, such as the Techni-Con Autoanalyzer.

13.4.6. Miscellaneous Ions

Almost any of the metal ions may be of interest in particular situations. Strontium, for example, is a minor component of sea water and of the carapaces of many marine animals, and may be a contaminant of interest in radiological studies. It, and many others, may be readily analyzed by atomic absorption spectrophotometry, provided that the proper lamps are available. A partial list of other ions that are amenable to atomic absorption analysis are arsenic, zinc, copper, cadmium, nickel, lead, barium, manganese, mercury, and chromium. Details on the analytical conditions for each are usually given in the instruments' manuals, but most require only suitable dilution in acid solution and a known set of standards.

Anions of various sorts are less often analyzed. Bicarbonate is a special case, discussed in connection with CO_2 analysis (Sections 5.2 and 10.4.2). Sulfate is most frequently determined gravimetrically by $BaSO_4$ precipitation and recovery. Nitrate and nitrite may be determined chemically (APHA, 1976) or, in the case of nitrite, with ion electrodes. For many ions, analysis kits are available from various companies that make otherwise tedious procedures quite painless. Reagent kits are available from Sigma, Hach, Boehringer-Mannheim, and others.

13.5. SPECIFIC ION ELECTRODES

A complete discussion of specific ion electrodes will not be attempted here, as several recent books deal exhaustively with the subject (e.g., Thomas, 1978; Sykova et al., 1981). In physiological work, only a few have found widespread application: the H^+ electrode (Chapter 6), the Na^+ electrode, the ammonia electrode, and, to a lesser extent, the K^+, Ca^{++}, and Cl^- electrodes. The pH, CO_2, and ammonia electrodes are all rather similar, except for the membrane/electrolyte systems employed, as discussed in various earlier sections. By manipulation of the composition of the soft soda-lime glass used for pH electrodes, electrodes selectively permeable to either Na^+ or K^+ may be produced, although the selectivity is not complete and the pH ranges over which they yield reliable results vary. There is some black art in making these, and they vary in quality from one manufacturer to another, so the specifications for a particular electrode should be examined closely. In general, they are suitable for situations in which the ion to be measured varies more than the interfering ions or those in which the interfering ions have a very low product of selectivity times concentration. The advantages of specific ion electrodes are that they are usually non-destructive and require little or no advance sample preparation. They do require larger amounts of sample than many of the methods discussed above, and in some cases pre-treatment, such as increasing the pH of the sample, is necessary, making the method destructive.

One important distinction in the use of ion-specific electrodes is that they sense *activity*, and not *concentration*. This concept has already been discussed (Chapters 2 and 6), and more detailed consideration of activity relationships in electrolyte solutions may be found in most chemistry texts. In biological systems, however, there are many circumstances in which the concentration and activity are very different, sometimes by several orders of magnitude, and no simple solution chemistry laws suffice to predict such relationships. The most outstanding example is calcium, whose intracellular activity may be three orders of magnitude less than the total intracellular calcium concentration. The difference is generally considered as "bound" ion, but the meaning of this term is not always entirely clear.

With a cation-selective electrode in a simple solution of the salt of the primary ion (M^+), the electrical response is Nernstian, that is to say, it obeys the equation

$$V_M = E_0 + (RT/ZF) \ln(aM^+) \qquad \text{(Eq. 13.3)}$$

where E_0 is a constant reference voltage, R the gas constant, T temperature in degrees Kelvin, Z the valence, F the Faraday, and aM^+ the activity of the ion in solution. If (13.3) is re-written with $\log_{10}$ instead of ln, it reduces to

$$V_M = E_0 + 58.1 \log(aM^+) \qquad \text{(Eq. 13.4)}$$

When there is a second interfering N^+ ion present, the selectivity constant K_{MN} is defined as the ratio of the change in electrode voltage produced by a unit change in the activity of the interfering ion divided by the voltage change produced by an equal change in the activity of the primary ion. The electrode's behavior may then be described by

$$V_M = E_0 + (RT/ZF) \ln[aM^+ + K_{MN}(aN^+)mn^{-1}] \quad \text{(Eq. 13.5)}$$

where m and n are the valencies of M and N. The selectivity of one type of Na^+ glass (NAS 11-18) is about 0.005, which means that a 1 mEquiv L^{-1} change in Na^+ activity produces 200-fold greater change in electrode potential than the same change in K^+ activity (Thomas, 1978).

The specific ion electrodes may generally be employed with any good pH meter, but most manufacturers offer pH meters modified to perform the log conversion and scaling required for direct concentration display ("pX" function). Some manufacturers of specific ion electrodes are Orion Research, Radiometer-Copenhagen, Corning, and HNU Systems. Extensive literature is available for any particular electrode and application.

13.5.1. Specific Ion Microelectrodes

In the past few years, many advances have been made in the development of microelectrodes for intracellular recording of ion activities. In the first designs, techniques similar to those used for pH microelectrodes were employed (see Chapter 6; Thomas, 1978), substituting the ion-selective glasses for the pH-sensitive glass. More recently, however, various liquid ion exchangers have become available, and have the advantages of ease of construction, better response and selectivity (in some cases), and much lower cost if labor is considered. There are now liquid ion exchange (LIX) formulations available for H^+, K^+, Ca^{++}, Cl^-, and others (see Sykova *et al.*, 1981).

13.6.   ION CHROMATOGRAPHY

High-performance liquid chromatography (HPLC) over the past few years has become the analytical method of choice for a large number of organic separation and identification procedures, most of which are outside the scope of this book. Lately, however, some new column materials and detectors have come on the market which are useful for analysis of a large variety of both anions and cations. In general, these methods appear to be more readily applicable to fluids which contain relatively low concentrations of ions, but with appropriate dilution and sample treatment, a large variety of physiological fluids could probably be analyzed by this method. The great advantage of the method is that a large number of ions are analyzed simultaneously by a single sample injection.

The principle is similar to that described for gas chromatography in Chapter 4. That is, ionic species to be analyzed are carried past a stationary phase by a mobile (solvent) phase. The ionic species are retarded to varying extents in passing through the column, and may then be detected sequentially upon exit from the column. The usual ion detector is a conductivity cell, but various other types, as well as fraction collection, are possible. Developments in this technology are rapid at present, so the interested reader is advised to obtain literature from one of the many companies offering equipment for HPLC and ion chromatography.

LITERATURE CITED

APHA. 1976. Standard Methods for the Examination of Water and Wastewater. 14th ed. Amer. Publ. Health Assoc., Washington, D.C.

Brown, N. L., & B. V. Hamon. 1961. An inductive salinometer. Deep-Sea Res. 8: 65–75.

Fiske, C. H., & Y. Subbarow. 1925. The colorimetric determination of phosphorus in body fluids. J. Biol. Chem. 66: 375–383.

Pollak, M. J. 1954. The use of electrical conductivity measurements for chlorinity determination. J. Mar. Res. 13: 228–231.

Sigma Chemical Co. 1981. The Colorimetric Determination of Inorganic Phosphorus in Serum and Urine. Bulletin #670. St. Louis, Mo.

Sigma Chemical Co. 1981. The Colorimetric Determination of Inorganic Phosphorus in Serum and Urine. Bulletin #670.

Sykova, E., P. Hnik, & L. Vyklicky (eds.). 1981. Ion-Selective Microelectrodes and Their Use in Excitable Tissue. Plenum Press, New York. 369 pp. (This is a symposium volume and has papers on many related topics.)

Thomas, R. C. 1978. Ion-Sensitive Intracellular Microelectrodes: How to Make and Use Them. Academic Press, London. 110 pp.

CHAPTER 14

RADIOISOTOPE TECHNIQUES

14.1. TYPES OF RADIATION

Atomic nuclei consist of protons and neutrons in roughly a 1 to 1 proportion, with increasing scatter about this ratio as the elements get heavier. The chemical nature of the element, of course, is determined by the number of protons and associated electrons in its outer shells, but for many elements there are several different numbers of neutrons that result in stable nuclear configurations. There are also combinations that are not stable, and these nuclei may undergo radioactive decay by several modes, all of which involve the emission of radiation of one or more types. Approximately 1500 unstable nuclear combinations, or *radioisotopes*, have been identified, with some elements having as many as 30 isotopes.

14.1.1. Alpha Decay

Some of the heavier isotopes have unstable combinations of neutrons and protons, and decay by emitting an alpha particle. The alpha particle is identical to the stripped helium atom, consisting of two neutrons and two protons. An example of this type of decay is found in the naturally occurring transformation series beginning with uranium:

$$^{238}U \rightarrow \,^{234}Th + \alpha \qquad \text{(Eq. 14.1)}$$

Most alpha particles have a narrow energy distribution around 4 to 6 meV. (The kinetic energy acquired by an electron accelerated through a potential of 1 V is called an *electron volt*, abbreviated eV, and is equal to 1.6×10^{-12} ergs.) For the reaction above, 77% of the alpha particles have an energy of 4.20 meV, and the daughter thorium nuclei are formed in their ground, or unexcited, state. The balance of the alpha particles has an energy of 4.15 meV; the thorium nuclei thus formed are in an excited state, and subsequently decay by one of two modes, releasing the remaining 0.05 meV of energy to reach the ground state.

Alpha-emitting isotopes are rarely used in physiological studies, although the effects of alpha radiation may be important in health studies.

14.1.2. Beta Decay

A large class of radioisotopes contain too many neutrons for the number of protons; they decay by "conversion" of a neutron to a proton by emission of an electron and a neutrino. The "daughter" nucleus, then, has the same atomic mass as the parent, but an atomic number one greater than the parent, as is the case for

$$^{14}C \rightarrow {}^{14}N + \beta^- \qquad \text{(Eq. 14.2)}$$

where the atomic number for the parent C is 6, and for the daughter N, 7. The distribution of energy between the electron (denoted as β^-) and the neutrino is not fixed, but is a stochastic process which results in a continuous energy spectrum like that shown in Fig. 14.1.

In certain nuclei with a low neutron-to-proton ratio, another mode of decay may occur in which a proton is transformed into a neutron with emission of a positron (denoted as β^+) and an antineutrino. An example of this type of decay is

$$^{22}Na \rightarrow {}^{22}Ne + \beta^+ \qquad \text{(Eq. 14.3)}$$

The atomic mass is not changed significantly by this decay mode either, but the atomic number of the daughter species is decreased, in this example from 11 to 10. As with β^- decay, the energy distribution of emitted positrons forms a broad spectrum, with the balance of energy taken by the antineutrino.

14.1.3. Gamma Decay

In any of the decay modes mentioned above, the daughter nuclei may be formed initially in excited, or more energetic, states than the normal rest state. These excited nuclei normally decay almost instantaneously to the ground state, emitting the excess energy as electromagnetic energy, or gamma rays.

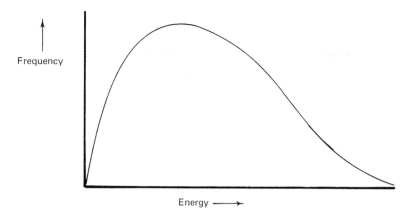

Fig. 14.1 An ideal beta energy spectrum, illustrating the continuous nature of the energy spectrum.

Gamma ray photons emitted by such decay have characteristically sharp energy distributions and may be highly energetic.

A second source of gamma rays which is of considerable practical significance in radioisotope applications is positron annihilation. The positrons emitted by β^+ decay will travel some distance until they start to lose energy, at which point they will combine with a normal electron. The resulting annihilation produces two gamma rays with energy of 0.511 meV, given off 180° in opposition to one another. In the case of ^{22}Na given above (Eq. 14.3), the annihilation gamma radiation is the principal type of radiation detected.

14.2. DETECTION OF RADIATION: INTERACTION WITH MATTER

The nature of the interactions between emitted radiation of various kinds and matter through which they pass is of importance both in detecting the radiation and in assessing its biological effects. The interactions of interest occur between the particles or photons emitted and the electrons in the orbital shells of atoms. The interaction may result in excitation of the electrons, which then leads to secondary radiation of light or other forms of energy; or it may result in ionization in cases where the electron is knocked free of the atom. Ionization is the principal result of the interaction of alpha and beta particles with matter, and the characteristics of the ionization may be used to detect these particles.

On the average, production of an ion pair requires about 35 eV of energy, so for a particle with an initial energy of 1 meV, roughly 30,000 ion pairs are formed before the particle reaches its rest state. For a given particle of a certain initial energy, the relationship between the distance traveled in various materials and the number of ion pairs formed is quite predictable. As the particle slows down toward the end of its travel, the number of ion pairs formed per unit distance rises, according to the curve shown in Fig. 14.2. The total range of travel is a function of the initial energy, the mass and charge of the particle, and the nature of the material through which it passes.

The maximum range of beta particles is shown in Fig. 14.3 as a function of the particle energy for several different materials. Also shown on the graph is the maximum beta energy for ^{14}C, a common isotope used in physiological studies. Complete absorption of this isotope in water is achieved in about 0.05 cm.

The interaction of gamma photons with matter may take three different forms, only two of which are of much significance in physiology. The first is the so-called photoelectric effect, in which the total energy of the gamma particle is absorbed by an atom, which reacts by ejecting an orbital electron. The energy of the orbital electron is equal to the initial gamma energy minus

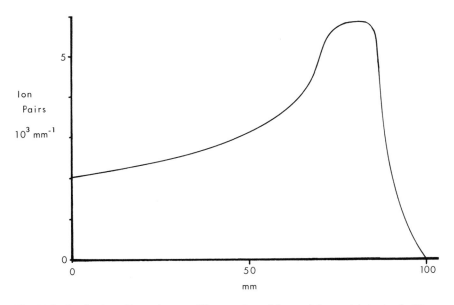

Fig. 14.2 Production of ion pairs per millimeter of travel for an alpha particle in air; the "Bragg curve."

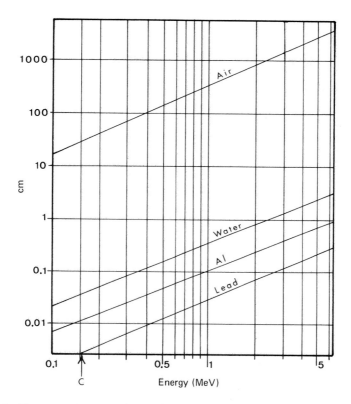

Fig. 14.3 The range of travel of beta radiation as a function of the energy of the radiation and the thickness of various materials. The arrow at the bottom indicates the beta energy of ^{14}C. (Modified from Wang, 1969.)

the binding energy of the electron. Since the energy distribution of the gamma rays forms a sharp peak characteristic of the radioisotope, the electrons emitted as a result of photoelectric interaction also have a sharp peak energy distribution. The second mode, particularly important for more energetic gamma photons or absorbing materials of lower atomic number, is the Compton interaction process, which produces so-called Compton electrons. In this type of interaction, the photon imparts only a part of its energy to an electron, and the initial gamma energy is divided stochastically between the excited electron and the gamma ray. The electrons produced in this way have an energy spectrum that is continuous, with an upper cut-off called the *Compton edge*. In Fig. 14.4 the absorption of gamma energy by both mechanisms is shown for aluminum, illustrating the predominance of photoelectric absorption at lower energies and Compton scattering at higher energies.

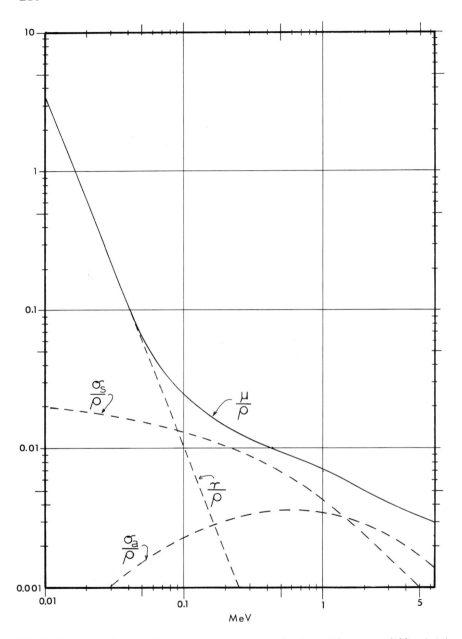

Fig. 14.4 The total attenuation of gamma radiation in aluminum. The upper solid line (μ/ϱ) indicates the total attenuation, which is the sum of photoelectric absorption (τ/ϱ), predominant at low energies; Compton scattering (σ_s/ϱ); formation of Compton electrons (σ_a/ϱ); and pair formation, which occurs at energies above those shown. (From data in Chackett, 1981, and Overman & Clark, 1960.)

14.2.1. Ionization Detectors

Since the interaction of various types of radiation with matter involves ionization, an obvious tactic for detection and measurement of it is to measure the ions produced. The earliest devices developed took the general form shown in Fig. 14.5, and consisted of an ionization chamber and an electrode system used to collected ions formed by interactions between the material in the chamber and radiation entering it. In the example shown, the center wire bears a positive charge and the outer container wall a negative charge. Ions produced by radiation will migrate according to their charge to the opposite electrodes, where an electrical current will be generated. The behavior of this type of detector is strongly dependent upon the charge applied between the electrodes, as the following discussion will show.

At a low applied potential difference between the two electrodes, many of the ion pairs formed will recombine with others in the chamber material (generally a gas) before reaching the electrodes, and so the detector will record fewer ion pairs than are produced. This is region A in Fig. 14.6. As the potential is increased, a plateau voltage is reached (B) at which all the ion pairs formed reach the electrodes, and the detector collects all of the formed ion pairs. Due to their larger mass and double charge, alpha particles are much more efficient at producing ion pairs, giving rise to the very different curves shown in Fig. 14.6.

With further increases in voltage above the plateau region, the primary ions formed by interactions are accelerated so much that they cause secondary ion formation by energetic collisions with other molecules of the gas in the chamber. The detector is acting in this region as an amplifier, and over a considerable range (Fig. 14.6D), the charge collected by the detector is greater

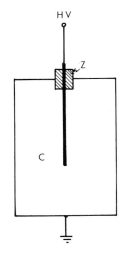

Fig. 14.5 Schematic diagram of a simple ionization chamber radiation detector. A high voltage (HV) is applied to a cathode in an evacuated chamber (C). The cathode is insulated (Z) from the wall, which serves as an anode and a sink.

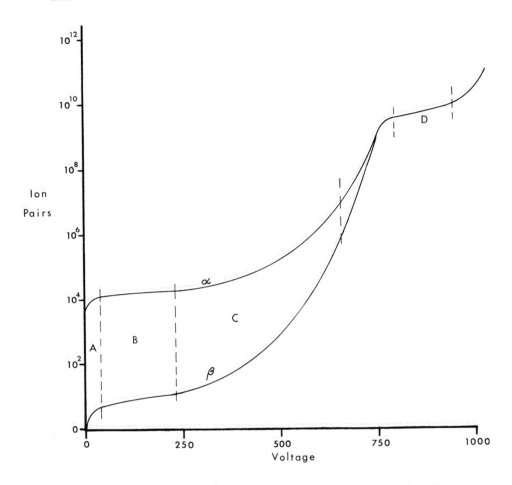

Fig. 14.6 A plot of the voltage applied to an ionization detector and the number of ion pairs formed. The different operating regions are defined and discussed in the text. The two curves shown are for alpha and beta radiation. (Modified from Overman and Clark, 1960.)

than but proportional to the number of primary ions. A detector operated in this mode is called a *proportional* detector, and is much more efficient as an alpha detector than as a beta detector for the same reason as given for the lower voltage region, i.e., a greater number of primary ions formed.

Finally, at quite high voltages, a region is reached in which every interaction produces a saturation response of the chamber. This is because the primary ions formed are accelerated so vigorously that they cause an "avalanche" of secondary ionizations. In this region (Fig. 14.6D), alpha and beta particles are counted alike, and the detector is called a *Geiger–Müller* detector after its inventors. Detectors based upon this idea are in fairly common use to detect total numbers of radioactive particles without regard to their type or energy. They are particularly common for safety monitoring of radioisotope work areas. For other physiological work, detectors based upon ionization chambers have largely been supplanted by other types, discussed below.

14.2.2. Scintillation Detectors

Solid Scintillators. A number of substances are now known which have the useful property of absorbing the excitation energy produced by interactions with radiation products and re-emitting part of that energy as visible light. Such materials are called *scintillators* and the process is known as *scintillation*. The earliest known scintillator was ZnS, but since this is not transparent to light, it can only be used in thin films and has been supplanted by transparent scintillators. These may be plastics, particularly anthracene-doped polystyrene, but are most often NaI crystals doped with thallium (Tl). The NaI(Tl) crystal is the most common type due to its high density, high capture efficiency, and suitable refractive index.

A typical NaI(Tl) crystal scintillation detector is shown in cross section in Fig. 14.7. Since the crystal is hygroscopic, it must be encased in a protective material. The crystal is usually covered on three sides by a very thin aluminum can which is coated on the inside with a reflective material and sealed on

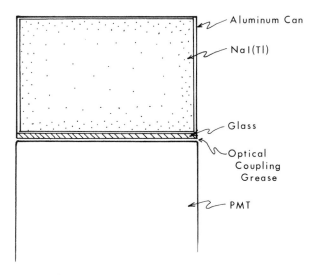

Fig. 14.7 Cross-sectional diagram of an NaI(Tl) crystal detector, showing the thin aluminum casing, and the connection to the photomultiplier tubes (PMT) via glass and optical coupling grease.

the fourth side with glass or plastic. The transparent bottom cover is mated with an optical coupling grease (to reduce reflection and refraction) to a photomultiplier tube for detecting the light pulses, and the whole assembly is mounted in a lead shield to reduce stray background radiation.

An ideal gamma spectrum detected by the crystal detector above would appear like the plot in Fig. 14.8A, with a sharp peak resulting from the photoelectric interactions and a lower-energy spectrum resulting from Compton electrons. In a real detector, the spectrum appears like that in Fig. 14.8B, where the ideal spectrum is scattered by a variety of non-ideal characteristics of the detectors and the interactions in real materials. The sharpness of the characteristic peak and the relative intensity of the Compton spectrum are partly a function of the size and geometry of the crystal. Larger crystals have much better resolution, and the geometry is improved by the "well" configuration, since it tends to equalize the path of travel for gamma rays emitted in different directions.

Solid scintillators may also be used for detection of beta radiation, but seldom are, since the penetration of beta particles in solid materials is quite small (Fig. 14.3). In some special applications, plastic scintillators are used in configurations that maximize the contact area between the scintillator and the sample and minimize the distance over which the beta particles must move. In most applications, however, liquid scintillation is preferred (see below).

Gamma counter design is a compromise between the resolution required for a given application and the cost of the crystal and shielding. For very low-level work, a large, well-shielded crystal is needed, but the cost of a 12.5 × 12.5 cm crystal may exceed $10,000. One figure of merit for a counter is the resolution, usually expressed for a particular isotope's principal peak as the width of the peak at half height as a percentage of the energy scale. For single isotope counting, 8–10% is fully adequate, but in experiments where two or more isotopes are counted simultaneously, the higher the resolution, the better.

Liquid Scintillators.   As shown in Fig. 14.3, the range of weak beta emitters in most solid materials is very small, especially for the commonly employed isotopes of carbon and hydrogen. The discovery of a number of liquids that scintillate has led to the design and wide application of the liquid scintillation counter, shown in schematic form in Fig. 14.9. The idea is to obtain a uniform dispersal of the radioactive material in an appropriate solvent for the scintillating materials (*fluors*), so that interactions may take place over molecular distances. The pulse intensity spectrum of the detected light is proportional to the energy spectrum of the beta radiation, which, as shown in Fig. 14.1, is continuous and non-linear.

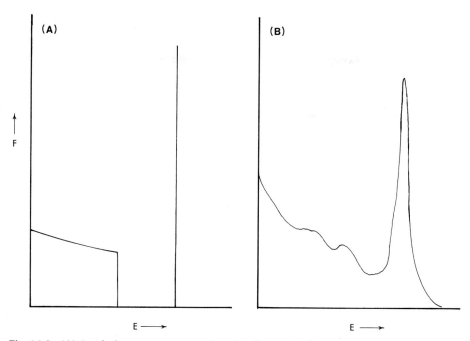

Fig. 14.8 (A) An ideal gamma spectrum, plotted as frequency of events (F) vs. the energy of each event (E), showing the total (photoelectric) energy peak and the Compton spectrum and edge. (B) An actual gamma spectrum.

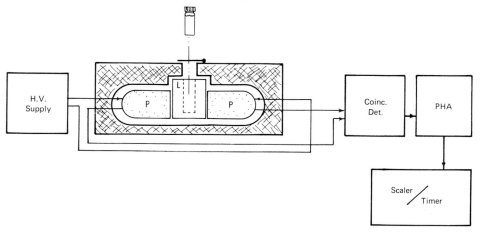

Fig. 14.9 A functional block diagram of a liquid scintillation counter, showing the high voltage (H.V.) supply, the pair of photomultiplier tubes (PMT) opposed to the lucite light pipe (L), and the electronic data processing modules.

14.3. RADIOACTIVE HALF-LIFE

Although many nuclear combinations of protons and neutrons are inherently unstable, poised for decay, as it were, it is a peculiar fact that they do not all do so at once. Rather, the decay process follows a probability function, such that in any given interval a reasonably predictable fraction of the nuclei will decay. The decay process, then, may be described by the exponential

$$f(t) = f(0)e^{-kt} \qquad\qquad \text{(Eq. 14.4)}$$

where $f(t)$ is the amount remaining at time t, $f(0)$ the initial amount, and k a rate constant. If the fraction of the initial amount at time t is designated as F, then

$$F = e^{-kt} \qquad\qquad \text{(Eq. 14.5)}$$

When F is exactly one-half, (14.5) may be reduced to

$$t = 0.693/k \qquad\qquad \text{(Eq. 14.6)}$$

and the value of t such that F equals one-half is called the *half-life* for the decay process. If the half-life for a particular isotope is given, (14.5) and (14.6) may be used to calculate the fraction remaining at any given time. For example, ^{22}Na has a half-life of 2.58 years, so the value of k from (14.6) is 0.269 and the fraction remaining after 4 years would be $e^{-4*(0.269)}$ or 0.341 of the initial amount. The graph in Fig. 14.10 may also be used to find the fractional amount of radioisotope remaining after any given number of half-lives. The half-lives of some radioisotopes commonly used in physiology are given in Table 14.1, along with information on their decay mode.

14.4. UNITS OF RADIATION

In order to have a convenient means of describing quantities of radiation, a variety of different units have been defined. Unhappily, these units are a source of great confusion, and the situation has not been aided by the constant tampering with units undertaken by the international bodies that consecrate unit conventions (such as the SI system). The various units must be carefully divided into those describing *radioactivity*, those relating to *exposure* to radioactivity, and those describing the *dose* of radioactivity received by a body.

The earlier unit of radioactivity was the Curie (Ci), defined as that amount of a radioisotope that yielded 3.7×10^{10} disintegrations per second (dps). This is the number (within 0.5%) of events occurring in 1 g of radium. The

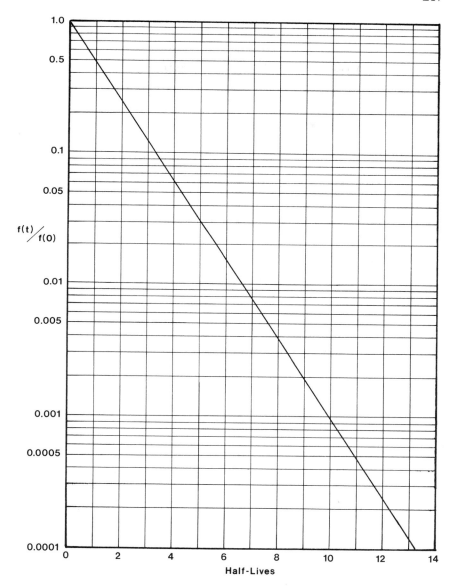

Fig. 14.10 A plot of the radioactivity remaining, $f(t)/f(0)$, as a function of the number of half-lives elapsed.

SI unit for describing radioactivity is the Becquerel, named for Henri Becquerel, who discovered radiation in 1896. The Becquerel, abbreviated Bq, is defined as that quantity of isotope which yields one event per second, and so has units $\sec^{-1}$. The Bq is inconveniently small; 1 megaBecquerel

TABLE 14.1
Properties of Some Radioisotopes Commonly Used in Physiological Studies [a]

Isotope	Half-life[b]	Specific activity (Ci/g)	Gamma energy (meV)	Beta energy (meV)
^{3}H	12.3y	9.8 × 10^3		0.018
^{14}C	5.73 × 10^{3}y	4.6		0.155
^{22}Na	2.58y	6.0 × 10^3	0.511, 1.27	0.55 (β^+)
^{24}Na	15.0h	9.0 × 10^6	1.37, 2.75	1.4
^{28}Mg	21.3h	5.3 × 10^6	1.35	0.46
^{32}P	14.3d	3.0 × 10^5		1.70
^{35}S	86.7d	4.0 × 10^4		0.17
^{36}Cl	3.0 × 10^{5}y	0.033		0.71
^{42}K	12.4h	6.0 × 10^6	1.5	2.0, 3.5
^{45}Ca	165.d	1.8 × 10^4		0.25
^{51}Cr	27.8d	9.0 × 10^4	0.32	(0.3)
^{54}Mn	303.d	8.0 × 10^3	0.83	(0.83)
^{59}Fe	45.d	5.0 × 10^4	1.1, 1.3	0.48, 1.57
^{60}Co	5.26y	1.1 × 10^3	1.17, 1.33	0.31, 1.5
^{63}Ni	92.y	6.0 × 10^2		0.067
^{65}Zn	245.d	8.0 × 10^3	1.12	0.33 (β^+)
^{86}Rb	18.7d	9.0 × 10^4	1.1	0.7, 1.8
^{90}Sr	28.y	1.44 × 10^2		0.6, 2.2
^{125}I	60.2d	1.7 × 10^4	0.035	0.03
^{131}I	8.05d	1.2 × 10^5	0.36	0.61
^{195}Au	183.d	3.6 × 10^3	0.1	0.09

[a]Data from Wang (1969).
[b]h = hours, d = days, y = years.

(MBq) corresponds to only 27 μg of radium. There is also a theoretical objection, namely, that radioisotopes do not decay on a deterministic basis; rather, they decay on a stochastic basis, so the Bq is definable only in an average sense, and not in absolute terms.

Exposure to radiation may be measured in various ways. Depending upon the particular isotope's decay scheme, a variable number of particles with a variable total energy may be emitted for each decay event. The product of the total number of particles and the total energy per disintegration represents the total radiant energy to which the surrounding environment is *exposed*. The primary effect of this exposure is the production of ion pairs, and it is usually easier to measure the production of ion pairs than to measure the number and energy of all emitted particles. The unit of exposure is the *Roentgen* (R), which is defined in a rather complex manner. It is the amount of radiation which will produce 2.58 Coulombs (C) of charge (ions of both signs counted) in 1 kg of air at STPD.

To express exposure in energy-related terms, we need the following calculations: 1 R produces 2.58×10^{-4} ion pairs per kilogram, each ion having a charge of 1.6×10^{-19} C, and each ion pair requires 34 eV of energy to produce. The unit of absorbed radiation energy, the *rad*, is defined by

$$(2.58 \times 10^{-4})(34)/(1.6 \times 10^{-19}) = 0.0088 \text{ J kg}^{-1} \quad \text{(Eq. 14.7)}$$

In the SI system, the absorption of 1 joule (J) kg^{-1} is defined as 1 *Gray* (Gy), so the Gray equals approximately 100 rads.

In spite of somewhat clumsy units and conversions, the physical description of radiation energy and absorption is accomplished precisely by the units listed above. In biological systems, however, the effects of absorbed radiation depend on more than the total physical energy absorbed. Such factors as the density of ionization along the particle track, the time during which the dose is absorbed, etc., make a considerable difference, and for purposes of comparing doses of different sorts of radiation a further unit, called the *dose equivalent*, is used. The unit of the dose equivalent is the *Sievert* (Sv), and is defined by

$$H = DQN \quad \text{(Eq. 14.8)}$$

where H is the dose equivalent in J kg^{-1}, D a general dose in Gy, Q an index of effectiveness of different types of radiation in biological tissues, and N a factor based on time fractionation and other variables. N is usually taken as 1, and Q is taken as 10 for neutrons and 20 for alpha particles.

14.5. PRACTICAL ASPECTS OF GAMMA COUNTING

14.5.1. Background Shielding

Natural radiation of a broad spectral range is constantly present, originating from sources as remote as deep space (cosmic rays) and as near as the materials used to construct the detector. In practical counting, a number of measures may be taken to reduce the background noise, and so increase the sensitivity and accuracy with which sample radiation is counted. The influence of crystal design and geometry on reducing Compton scattering in the low energy range has already been discussed. Heavy shielding of the detector with lead is also standard procedure, although the amount of lead supplied in various commercial detectors varies greatly, depending on the price and the intended application. Another often neglected source of extraneous radiation is other samples waiting to be counted. If at all possible, especially with high energy emitters, other samples should be moved as far from the counter as possible. As with radiated light, the intensity of radiation encountered per unit area decreases as the inverse square of the distance, so compromises are possible beyond a reasonable distance. In multi-sample counters this is not possible, and one must rely on the shielding provided by the manufacturer.

14.5.2. Efficiency Correction

The total number of counts detected by any given counter is a function of the sample radioactivity, the energy of the emitted radiation, and the counter efficiency. It is particularly important to know the efficiency if more than one counting device is employed, and this is best measured on a relative basis by counting a calibrated standard, which may be supplied by one of several companies. Even with a given counter and detector, however, the counting efficiency may vary depending on the geometry of the sample and the nature of the material in which the radioisotope is dispersed. Obviously a sample of ^{22}Na will not count with the same efficiency when dispersed in a soil sample as when coated on the wall of the sample container. Also, a sample vial will have a different geometry depending on how full it is (Fig. 14.11). Since it is usually important to compare counts in a series of samples, one should try to ensure that each sample container is filled to about the same extent and that the samples are not dispersed in widely different materials. Counting efficiency for most gamma counter systems for medium-energy gamma radiation is relatively low, in the range of 5 to 15%.

14.5.3. Statistical Counting Error

Although on the average the decay of radioactive isotopes follows the exponential function of Eq. 14.4, in any given (small) interval of time, the counts will fluctuate with a random element determined by probability. As a rough rule of thumb, the standard deviation of the total counts is approximately equal to the square root of the total counts. That is, if only 100 counts are recorded in a given interval, the standard deviation to be expected would be nearly 10% of the estimated count. To improve this to 1%, 10,000 counts would have to be recorded, although the time of counting also enters into the calculation. The nomogram given in Appendix 6 provides a quick way to calculate the 90 and 95% confidence intervals for various counting rates and times.

When count rates are low, approaching the background activity, the square root rule applies to both the combined sample + background count and the background count alone. The standard deviations of the difference, which is usually the measure of interest, can be estimated by the formula

$$S_{s\text{-}b} = (C_s/T_s + C_b/T_b)^{1/2} \qquad\qquad (\text{Eq. } 14.9)$$

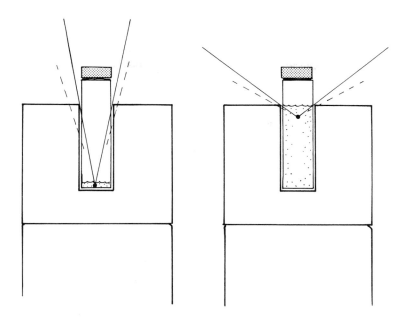

Fig. 14.11 Diagram of a well counter to illustrate the effect of sample volume on counting efficiency. In the example on the right, a greater angular loss will result from the geometry of the system, and the length of the average path of gamma particles through the crystal will also be less, on the average.

where C stands for the counting rate over T minutes, and s and b stand for sample and background, respectively.

14.5.4. Multi-Isotope Counting

The overlapping spectra shown in Fig. 14.12 for ^{59}Fe and ^{51}Cr would produce the combined spectrum shown, i.e., the sum of the two. The calculation of each isotope's activity is performed by correcting for the overlap of each isotope in the "window" of the other. In the example shown, for regions A and B, the counts in A consist of a contribution from each isotope. The distribution of each isotope in each energy region must be determined by counting known standards of each alone, which allows calculation of channel ratios. These ratios are then used in a series of equations which may be solved for the total counts due to each isotope alone. A full derivation of these equations is given in Appendix 7. The calculations for more than two isotopes get quite complicated and generally require computer assistance to solve, but the principles are the same, and calibration of the system with

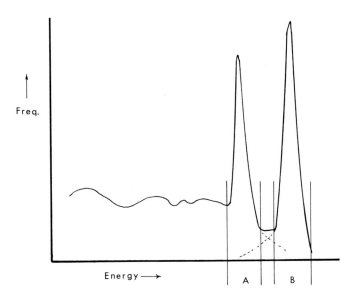

Fig. 14.12 Diagram of the gamma spectrum for two isotopes, ^{51}Cr (A) and ^{137}Cs (B), showing the regions of overlap, A and B. Calculations for this case are developed in Appendix 7.

individual standards is necessary. If the energy spectra of the isotopes are close to each other, crystal resolution becomes critical and the error of the final results increases.

One neat way of counting more than one isotope is to select isotopes with quite different half-lives, one very short and the other relatively long. One can then count the sample activity due to both isotopes, wait a sufficient period of time to allow decay of most of the shorter-lived isotope, and then count the sample again. The difference between the two counts will be the activity due to one isotope, and the second count the other. ^{24}Na and ^{36}Cl are two common isotopes that may be counted together in this way; the ^{24}Na decays in a few days, leaving only the ^{36}Cl activity for the second count.

14.6. PRACTICAL ASPECTS
OF LIQUID SCINTILLATION COUNTING

14.6.1. The Counter

Since many applications of liquid scintillation counting are concerned with quite low levels of radioactivity, several design improvements have been incorporated into most commercial instruments to reduce the background count rate. In Fig. 14.9, a cross-sectional diagram shows the arrangement of the sample well in a lucite light pipe. This light pipe is coupled by an optical coupling grease to two opposed photomultiplier (PM) tubes, and the whole assembly is encased in a heavy lead shield. The entry path for the sample holder is closed off by a light-tight shutter which may also be shielded.

The common sources of noise include stray light, which is easy to eliminate; background radiation including cosmic sources, which can largely, but not completely, be eliminated; thermionic noise from the photomultiplier tubes; and stray emission from the sample. In early instruments the detector assembly was usually kept refrigerated, partly to reduce PM tube noise. Newer tubes are inherently quieter, and most of the remaining noise is canceled out by the electronic processing of the signals from the two opposed tubes. A flash of light (scintillation) in the sample should be observed simultaneously by both PM tubes, whereas stray thermionic noise is likely to occur asynchronously in the two. Therefore, by operating the two tubes in a coincidence mode, in which only simultaneous electrical pulses from the two PM tubes are counted, most of the remaining noise is eliminated. A good counter may have a background counting rate of only 15 to 25 counts per minute, even in the low energy tritium region.

A further reason for refrigerating the detector used to stem from the nature of the fluors used (see below). Except for special purposes, it is probably best not to refrigerate the detector. Keeping it cool causes condensation, and this may lead to problems of efficiency and repair.

14.6.2. Scintillation "Cocktails"

Samples are usually prepared for liquid scintillation counting by adding them to a counting "cocktail" which is made up of several elements. The organic *fluors* serve to absorb electron energy produced by the sample radiation and re-emit part of the energy as visible light. The most commonly used such compound is 2,5-diphenyloxazole; its structure is shown in Fig. 14.13, and it is usually known by the acronym PPO. The fluorescence wavelength peak for this compound is 365 nm, however, which is somewhat outside the peak absorption efficiency range of most PM tubes. A secondary fluor, then, is often used in concert with PPO; the most common such fluor is 1,4-bis-2-(5-phenyloxazoyl)-benzene, so-called POPOP, or its dimethyl derivative. These have fluorescence maxima at 418 and 429 nm, respectively, providing better PM tube capture efficiency.

The fluors in most cases are non-polar compounds that are not soluble in water, so a solvent system must be used that will provide a single-phase solution of both the fluors and the sample and will not interfere with the scintillation counting. Most commercial cocktails have been based upon either xylene or toluene as the primary organic solvent, but recently many companies

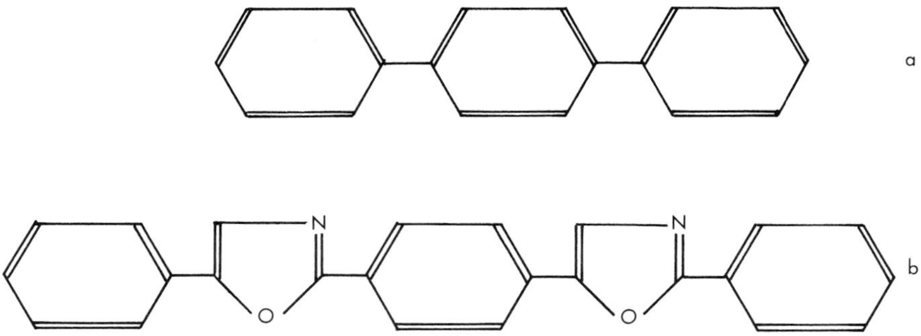

Fig. 14.13   The chemical structure of two organic fluors for liquid scintillation: (a) para-terphenyl and (b) POPOP.

have shifted to solvents with higher flash points, mostly for safety reasons (see below). A wide variety of emulsifiers may also be used. One recipe for aqueous samples is the following:

PPO	4. g L^{-1}
POPOP	0.1 g L^{-1}
Toluene	1 L
+ Triton X-100	

with the amount of Triton X-100 varying up to 40% depending on the amount of water to be added to the final mixture. Many other combinations are in use, however, and one often finds that it is simpler and less expensive to take advantage of commercial cocktails designed for specific applications. Many early recipes were based upon dioxane as the primary organic solvent, and these cocktails had several bad habits. With substantial water content added, they had a tendency to separate into two-phase systems, which do not count properly, and the dioxane has considerable chemiluminescence after exposure to light. Samples prepared with dioxane cocktails, then, must be kept cool and in the dark for some period of time prior to counting.

14.6.3. Sample Preparation

The principal advantage of liquid scintillation counting is that it provides very short-range exposure of the fluors to the radioactive particles; in a true solution, distances would be sub-micron, providing a high probability of interaction with the fluor, as opposed to absorption in non-scintillating sample or solvent materials. One must therefore be quite careful to ensure that the sample material is really dissolved in the cocktail and not simply suspended in it as particles. For tritium counting, this is especially critical in view of the very short range of the weak β^- particles emitted (Fig. 14.3). For biological tissues, this usually means that some means of digestion or combustion should be employed. The most common means are chemical digestion with strong organic bases and complete combustion to CO_2 and H_2O in a sample oxidizer.

For single isotope work, say with ^{14}C or ^3H, chemical digestion is easier and does not require any expensive equipment. Various organic bases are available under trade names, such as Soluene-350 (Packard) and Protosol (New England Nuclear). Tissue samples are simply added to an aliquot of the base and allowed to digest before the cocktail is added. Two problems commonly occur, however, which require further treatment. The digests are often yellowish or, in the case of tissues containing blood residues, red. These colors interfere with the detection of scintillation from the fluor and must be bleached out with a peroxide treatment, which may cause secondary chemical quenching. The second common problem is that many cocktails exhibit either strong quenching (see below) or chemiluminescence with strong bases added, so that neutralization after digestion may also be required.

In order to avoid these problems, a much cleaner method is to oxidize (combust) the organic material in the samples and count it as the CO_2 or water produced. For ^{14}C a simple furnace oxidizer system (Fig. 14.14) is easily constructed; it provides high recovery efficiency and fairly rapid sample processing. The CO_2 is trapped in ethanolamine or related compounds sold under various trade names; then cocktail is added and the sample can be counted with virtually no color or chemiluminescence interference. Recovery of water from such a system is much more difficult, however, since the water has a tendency to condense on any cooled surface and acts as a "sticky" gas

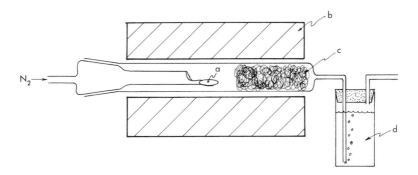

Fig. 14.14 A simple furnace (b) oxidizer system for collecting ^{14}C. A CO_2-free nitrogen stream (N_2) is passed over the sample holder (a) at about 600°C, and thence over a copper oxide catalyst (c) to ensure complete combustion. The resulting CO_2 is then trapped (d) in ethanolamine or a similar agent.

which is hard to collect. At present, the only satisfactory sample oxidizer system available commercially is the Packard Sample Oxidizer (Fig. 14.15). It provides semi-automated processing of samples containing both ^3H and ^{14}C, collecting each separately in an automatically mixed absorbent/cocktail system.

14.7. QUENCHING AND EFFICIENCY CORRECTION

14.7.1. Quenching

The ideal β^- spectrum shown in Fig. 14.1 is never actually obtained from a liquid scintillation counter. A number of intrinsic limitations set the efficiency at less than 100%, and quenching further reduces this efficiency. *Quenching* is a reduction in counting efficiency which may be caused in a variety of ways by materials which interfere with the translation of the energy of beta particles to visible light. The general effect of quenching, shown in Fig. 14.16, is to shift the observed energy spectrum to the left and downward to a variable degree. The use of a single term for this phenomenon is perhaps

Fig. 14.15 The Packard Sample Oxidizer, a device for combustion of organic materials to CO_2 and water and for separate collection of each in appropriate scintillation cocktails. The machine is designed for dual-label work with ^3H and ^{14}C.

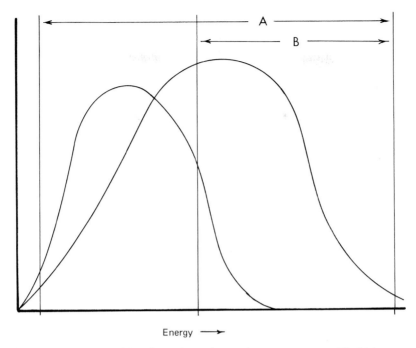

Energy ⟶

Fig. 14.16 An illustration of the effects of quenching on beta energy spectra. The higher energy curve to the right is unquenched, the attenuated, left-shifted one quenched. Adjusting two energy windows for counting to the regions shown as A and B may be used in one method of correction for quenching, the so-called sample channels ratio (SCR) method (see text).

unfortunate, since many different processes may contribute to quenching, and the proper corrections for each differ considerably.

14.7.2. Chemical Quenching

A wide variety of impurities in the sample or any part of the cocktail may interfere with the complex series of energy transfers at nearly any point. For example, binding between impurities and the fluors may prevent luminescence, the luminescence may be absorbed by the contaminants, the radiated electrons may directly interact with the impurities, and so forth. The net result in all cases is that part of the energy of the original radiation is lost, and some emitted particles may be lost altogether. The effect on the observed spectrum is to shift it toward lower energies (i.e., to the left in the conventional manner of display) and to reduce the total number of events observed, as shown in Fig. 14.16. Many common compounds, such as acetone, exhibit strong chemical quenching, but nearly all compounds have some quenching activity, even dissolved oxygen.

14.7.3. Color Quenching

Quite another kind of quenching arises from spectral interference of colored compounds with the light emitted by the fluors. For example, hemoglobin, a common tissue "contaminant," absorbs strongly in the wavelength region where the fluors luminesce, causing a severe shift in the observed spectrum even at very low concentrations. Many other proteins and biological compounds may contribute significantly to color quenching without being obvious upon visual inspection. In many cases color quenching can be alleviated by bleaching, but the bleaches may themselves introduce chemical quenching or produce chemiluminescence. An important point to emphasize is that the effect of color quenching on the observed energy spectrum is different from that produced by chemical quenching. While this may not be serious in many applications, it may occasionally cause very large errors if not recognized.

14.7.4. Quench Correction Methods

Internal Standardization. The principle of internal standardization is quite simple, and it provides the most reliable method for efficiency correction in the face of unknown or variable quenching. One simply counts the sample once, adds a precisely measured amount of a standard containing known radioactivity of the same isotope, and counts the sample again. Thus addition of 10,000 dpm of standard will produce a count 5000 dpm greater if the efficiency is 50%. Although superior in principle, this method has several drawbacks. First, the sample must be counted twice, with some intervening time interval. For short half-life isotopes, this may present a problem, and of course, an additional possibility of pipetting error is present when adding the calibration standard. If samples are being counted in plastic scintillation vials, solvent is lost through permeability of the vial, and this will change the efficiency from the first count to the last. Finally, considerable time and expense are involved in performing this standardization for each sample.

Sample Channels Ratio Correction. In most liquid scintillation counters, two or more channels are provided which may be set to count over different energy ranges. Since quenching shifts the spectrum toward the lower energy range, the two channels may be adjusted so that the range of quenching in the samples yields a workable range of ratios of counts in the two channels. To illustrate this, energy spectra for a lightly quenched and a heavily quenched sample are shown superimposed on two channel regions in Fig. 14.16. For the least-quenched sample, the ratio of A to B would be high and for the most-quenched one, low. The ratio of these two channel counts may be related to counting efficiency by counting a series of variably quenched standards and computing the sample channels ratio vs. efficiency for the series.

There are also pitfalls with this method. The series of quenched standards should ideally be of the same composition as the samples to be corrected. More often, a commercial series of quenched standards, made up in toluene with perhaps acetone as a quenching agent, is used to set up a sample channels ratio (SCR) correction for samples containing any of a variety of cocktails and quenching agents. If the samples are color quenched, a color quenching agent should be used for the calibration curve; if chemically quenched, a similar chemical agent should be used. Significant errors may result from ignoring the difference in the effect on the observed spectrum.

A recommended procedure is to make up a large batch (about 200 ml) of known standard radioisotope solution in the cocktail to be used for the samples. Aliquots of this are then placed into vials, say 10 vials of 18 ml each, and counted to 1% or less accuracy. One standard is then left untreated, and progressive amounts of quenching agent are added to the rest of the series, which are then re-counted and their SCRs computed. If known, the same quenching agent found in the samples should be used for this step. The second counting after addition of the quenching agent will then allow calculation of the counting efficiency vs. SCR relationship, and this relationship is then used to correct each sample to true dpm.

External Standard Ratio Correction. The third common method for efficiency correction employs an integral gamma source that is incorporated into most liquid scintillation counters. This gamma source, normally shielded from the detector, is mechanically brought close the sample vial, irradiating it with high-energy gamma rays (Fig. 14.17). These gamma rays produce Compton electrons whose energy is generally well above that of the β^- particles to be counted, and the counter's electronics automatically shift to two energy bands above the normal range. Since quenching will usually produce the same effects on the Compton spectrum as on the beta spectrum, the channels ratio from this external standard may be used with a series of quenched standards to construct an efficiency correction curve in much the same manner as with the SCR method. This method is superior for many applications but is sensitive to fluor volume, so the volume of standards and samples should be the same, if possible. There may also be some differences between color quenching in samples and chemical quenching in standards, so, as with the other methods, this should probably be checked by internal standardization of at least some of the samples.

There is a natural tendency to ignore some of these rather subtle problems in liquid scintillation counting, since one always obtains a count of some sort, and the counts may appear to be reasonable. When one can be really

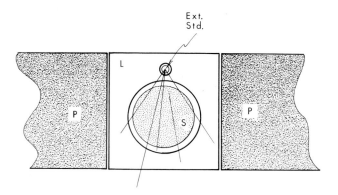

Fig. 14.17 The physical arrangement of the external standard pellet (Ext. Std.) in a liquid scintillation counter. The pellet is conducted from a shielded region through a tube so that it comes to lie next to the sample (S), irradiating it in a high energy region. Scintillations are transmitted through the lucite light pipe (L) to the photomultiplier tubes (P).

sure that all the samples are quenched to the same degree, many of these subtleties can be ignored, and relative activity calculations, using counts per minute rather than disintegrations per minute, may be used. Often, however, the problems are more serious than one realizes, and the importance of using standards with the same composition as the samples for quench and efficiency correction cannot be over-emphasized.

14.7.5. Dual Isotope Beta Counting

The complexities of counting two or more gamma isotopes (Section 14.4) are compounded further by quench and efficiency correction problems in liquid scintillation counting. The commonest pair counted is ^{3}H and ^{14}C, and the following discussion will pertain only to them. The usual procedure is to set the upper energy channel in such a way that ^{3}H is virtually excluded. This means that the contribution of ^{14}C to counts in the lower energy channel must be made, but not the reverse correction, which simplifies life. The general procedure is as follows: First, a series of ^{14}C standards must be used to establish two relationships: the counts in the upper channel vs. ^{14}C efficiency, and the ESR vs. ^{14}C contribution to counts in the lower channel (i.e., the SCR). Then a separate external standard ratio vs. ^{3}H efficiency curve must be established. When unknown samples are counted, the upper channel count may be simply converted to ^{14}C dpm. Then, the upper channel count is multiplied by 1/SCR and subtracted from the lower energy channel. Finally, the remainder from the low energy channel is corrected to the true dpm for tritium. With all of these calculations and manipulations, small errors in the relationships required may combine to produce rather large errors in the calculated dpm values, especially for tritium. Reducing the quenching and arranging the radioactivities in such a way as to have 4 to 10 times as much ^3H activity as ^{14}C help considerably.

14.8. RADIATION SAFETY AND HANDLING

Government and institutional rules spell out in detail the required procedures in the purchase, storage, and disposal of radioisotope materials, as well as record-keeping related to their use. Anyone contemplating radioisotope use should thoroughly familiarize himself with the federal and institutional regulations and obtain the necessary permits. Fortunately, nearly all physiological applications of radioisotope methodology require only very small quantities of radioactivity, so the exposure hazard is minimized. Nevertheless, most authorities feel that there is *no* absolutely safe exposure limit, so every effort should be made to avoid unnecessary exposure.

Anyone working with amounts of radioactivity above an approved minimum level will be required to wear some type of exposure monitor, the most common of which is the film badge. A typical film badge is shown in Fig. 14.18. The principle of this device is the original observation by the Curies, who discovered radium by its effect on photographic film. The film badge typically contains a light-shielded piece of special photographic film, portions of which are covered by two pieces of metal of different thickness. After some period of time, typically a month, the film is developed and the dark grains resulting from radioactive interaction with the film emulsion are counted. By comparing the number of exposed grains in the uncovered portion with the number under the metal layers, a rough measure of the penetration, and hence the energy of the radiation, is also obtained. In most institutions, regular surveys of laboratory surroundings are also conducted to measure the constant low-level exposure received by workers in those areas.

In addition to the institutional procedures, however, a number of precautions in laboratory procedures are in order. Pipetting should always be performed with the aid of a mechanical bulb or other device, not by mouth. Radioisotope work areas should be identified and kept separate, if possible, from the rest of the laboratory work areas. It is good procedure to cover laboratory benches with a plastic-backed, disposable paper mat, so that any spills may be easily controlled. Radioisotope solutions made up in bottles or flasks should always be carried in another container, preferably

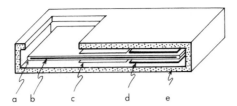

Fig. 14.18 A cross-sectional diagram of a typical film badge, showing sections of the film (b) covered in light-tight paper (a). Portions of the film are covered either by thin aluminum (c) or copper sheet (d) to provide an estimate of the energy of incident radiation. The whole assembly is packaged in a plastic carrier (e) that attaches conveniently to clothing.

unbreakable and lined with absorbent material. A cut-off plastic bottle with crumpled tissue paper makes an inexpensive and effective container. Stock solutions should be stored in a shielded area or should at least be moved as far away from regular work areas as possible. When materials are disposed of in the sink systems (observing maximum allowable limits set by the Nuclear Regulatory Commission), they should be poured carefully straight into the drain and then rinsed with copious quantities of water. When spills occur, the proper clean-up procedure will depend upon the particular isotope involved, the quantity of radioactivity, and the solvent. The procedure may be simple, as for a small spill on a non-absorbent surface, or in an extreme case may involve complete removal of structural materials into which the radioisotope material may have been absorbed. If there is any doubt, the institutional radiation safety people should be called in, and careful monitoring conducted afterward to ensure that the clean-up was complete. It is far easier to observe safe handling procedures in the first place than to clean up later.

The relative hazards of various radioisotopes vary, depending on a number of factors. Some of the obvious ones are the total radioactivity involved and the energy of its emitted radiation. Tritium, as a common example, has a sufficiently weak beta emission that an aqueous solution in a glass container is completely shielded by virtue of absorption in the water and the glass (cf. Fig. 14.3). ^{22}Na, on the other hand, has a relatively "hard" gamma emission that may be completely shielded only by thick lead. The dangers of accidentally ingesting material, however, may be inversely proportional to the radiation energy. Since the absorption of beta energy from ^3H is complete over such a short distance, this means that the ionizing events associated with the absorption will all take place in the tissue where the decay occurs. For a hard gamma emitter, a high proportion of the emitted radiation will pass through the tissue and produce no ionization or other interaction.

There is a further important consideration in assessing the hazard, and that is the biological half-life, or residence time. That is, the hazard is also a function of the length of time, on the average, that accidentally ingested or absorbed material will stay in the body before being excreted or lost in some way. Thus, ^{45}Ca is relatively dangerous due to its tendency to become incorporated into bone, thus having a nearly life-time residence. Iodine isotopes are similarly dangerous due to their sequestration in the thyroid gland. A further consideration is the radioactive half-life; naturally, shorter half-lives lead to a lower total exposure, since the activity does not persist for long periods of time.

A final hazard which should not be ignored is the chemical toxicity of the chemicals employed in liquid scintillation counting, especially toluene, xylene, and other aromatic solvents. The recent switch of many companies to higher flashpoint solvents for scintillation cocktails will reduce the fume concentrations in the laboratory, but caution is still indicated. In particular, solvents should be stored in well-ventilated areas, and used plastic scintillation vials should not be stored where the fumes will enter the working areas of the laboratory.

In all work with radioisotopes, zero contamination should be the target, and techniques and laboratory arrangements should be examined critically at all times to see whether that goal is being approached. Most of the precautionary measures are no more than good laboratory technique anyway. Radioisotopes in the tracer quantities employed can be handled safely, but they should always be treated as extremely hazardous materials.

LITERATURE CITED

Chackett, K. F. 1981. Radionuclide Technology. Van Nostrand Reinhold Co., New York. 426 pp.

Howard, P. L., & T. D. Trainer. 1980. Radionuclides in Clinical Chemistry. Little, Brown, Boston. 154 pp.

Overman, R. T., & H. M. Clark. 1960. Radioisotope Techniques. McGraw-Hill, New York. 476 pp.

Siegbahn, K. 1955. Beta- and Gamma-Ray Spectroscopy. North-Holland Publishing Co., Amsterdam. 959 pp.

Wang, Y. 1969. Handbook of Radioactive Nuclides. CRC Press, Cleveland. 960 pp.

CHAPTER 15

TRACER METHODS AND COMPARTMENTAL ANALYSIS

15.1. INTRODUCTION

The application of radioisotope tracers to physiological problems is perhaps one of the most powerful analytical techniques ever developed. Except for minor differences in mass, these tracers behave exactly like their natural counterpart compounds, yet can be easily detected and measured in minute quantities. The availability of radioisotopes of elements such as carbon shortly after World War II led almost immediately to fundamental new insights into photosynthetic pathways, biochemical pathways, and rates of many physiological processes. New applications of radioisotope technology continue to appear, extending into the clinical and even the consumer realm (e.g., smoke alarms). In physiology, tracer methods are commonly applied both for qualitative studies, such as pathway elucidation, and for quantitative studies of the rates of many processes.

A basic concept in tracer analysis is that of the *compartment*. In some cases, the compartment may correspond to a physically bounded space, such as the cerebrospinal fluid compartment or the blood plasma compartment. In many other cases, a pool of like molecules which tend to behave the same may be regarded as a compartment, even though the molecules may be heterogeneously distributed throughout the body. An example of the latter

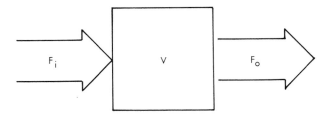

Fig. 15.1 Diagrammatic representation of a dye wash-out model, showing the volume (V), inflow (F_i), and outflow (F_o).

might be the body glycogen pool, which for some purposes may be modeled as a single compartment. Organisms tend to regulate their physiology in such a way as to maintain these pools or compartments at constant size, even when addition and removal are occurring at relatively high rates. Such a situation is termed a *steady state*, and the following analyses pertain exclusively to steady state situations.

15.2. SINGLE COMPARTMENT KINETICS: THE DYE WASH-OUT EXAMPLE

To use the simplest example, let us assume that we have a certain volume, V, of water, which has an inflow F_i and an outflow F_o which just balance each other, such that V remains constant (Fig. 15.1). In some circumstances, all of the desired information about this compartment may be obtained by direct measurement, for example by measuring the volume and the two flow rates. In many cases, however, direct measurement of some of the parameters is not possible, and tracer methods allow indirect measurement. Assume in our example of Fig. 15.1 that the only piece of information available is that V is constant, and that we wish to obtain measurements of the flow rates and the volume. A simple application of tracer methods will provide this information.

Assume at time t_0 that an amount of a non-reactive tracer material (such as polyethylene glycol) labeled with ^{14}C is added with instantaneous mixing to give a total radioactivity of 1 μCi (2.2×10^6 dpm) and that a series of samples is taken over the next hour at 5 min intervals. The data would appear as in Fig. 15.2. The rate of change of material (Q) in the compartment at any given time can be written as

$$dQ/dt = FQ/V \qquad \text{(Eq. 15.1)}$$

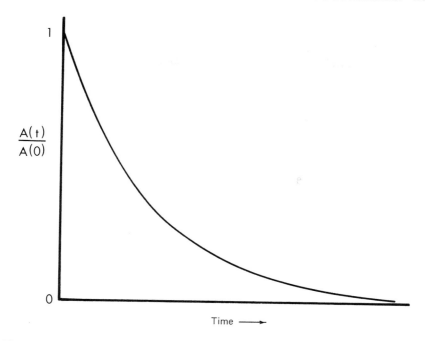

Fig. 15.2   Plot of the fraction of initial activity $(A(t)/A(0))$ of a tracer substance remaining in the compartment of Fig. 15.1 as a function of time.

and integrating this equation yields

$$Q(t) = Ae^{-(F/V)t} \qquad \text{(Eq. 15.2)}$$

where A is a constant from the integration. This first-order equation describes *exponential decay*, which has already been encountered in control of gases (Chapter 9) and in connection with radioactive decay (Chapter 14). If the data of Fig. 15.2 are plotted on a semi-log scale, the relevant equation (15.2) becomes

$$\ln Q(t) = \ln A - (F/V)t \qquad \text{(Eq. 15.3)}$$

and the re-plotted data appear as shown in Fig. 15.3. The zero time intercept corresponds to the value for A and the slope to the *rate constant, k*. For the data shown in Fig. 15.3, the slope is 0.1 and the zero intercept value of A must be simply the total quantity of radioisotope added. We can easily calculate that $k = F/V = 0.1$, and Eq. (15.2) becomes

$$Q(t) = (2.2 \times 10^6)e^{-t/10} \qquad \text{(Eq. 15.4)}$$

This example may also be put into concentration terms by dividing the

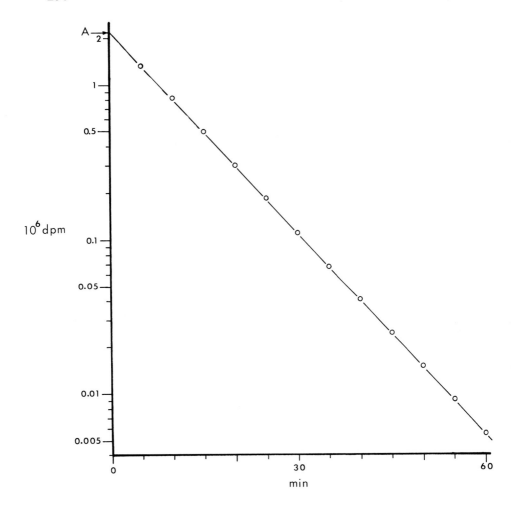

Fig. 15.3   A semi-log plot of the same data as in Fig. 15.2, using the values given in the text example. The ordinate scale is in units of total radioactivity but could just as well be measured and plotted in concentration units (see text).

quantity in the pool Q by the volume V. If our measurements had been made in terms of concentration rather than total quantity, the scale of Fig. 15.3 would be in units of Q/V, or concentration. The zero intercept then gives the zero time concentration, which is the total quantity added divided by the volume. Supposing our data showed a zero time intercept of 2.2×10^4; we could then calculate a total volume V of 100 ml, and so from Eq. (15.4) one obtains

$$C = (2.2 \times 10^4)e^{-t/10} \qquad \text{(Eq. 15.5)}$$

and the original equation describing the loss would be

$$dC/dt = FC/V \qquad \text{(Eq. 15.6)}$$

One further useful rearrangement of these equations is to express the amount or concentration remaining at any given time as a fraction (S) of the initial quantity, giving

$$C_t/C_0 = S = e^{-kt} \qquad \text{(Eq. 15.7)}$$

In applying this analysis, one should be aware of the assumptions implicit in the ideal case. One is that the indicator substance becomes instantaneously and uniformly mixed in the entire pool. In practice this will never be so, but as long as the mixing time is short relative to the rate of loss, violation of this assumption will not cause a very serious error.

We have now arrived at the information desired at the outset. From the zero-time extrapolated radioactivity we calculated the size of the compartment:

$$V = D/Q_0 \qquad \text{(Eq. 15.8)}$$

where D was the total radioactivity added. By fitting a regression line to the semi-log plot of the data in Figs. 15.2 and 15.3, we derive the slope, k. From the slope we can calculate the material flows by

$$F_i = F_o = kV \qquad \text{(Eq. 15.9)}$$

and the compartment is completely characterized.

15.3. SINGLE COMPARTMENT ANALYSIS: A CHEMICAL POOL EXAMPLE

In the above example, fluid flow was the mechanism of movement of material into and out of the compartment, and the compartment itself was a physical space and volume. For that example, a dye could just as well have been used to assess flow and volume. More often the compartment is chemical in nature, and the size of the compartment is even more difficult to measure directly. As a further example of single compartment kinetics, the behavior of creatinine in mammalian blood plasma serves as a good example. Creatinine is the end product of phosphocreatine metabolism, and is excreted by filtration in the kidney according to the model in Fig. 15.4. It may thus be modeled as a single compartment with a single exit and a single entry pathway, represented by k_1 and k_2. For this example, it is more convenient to work in terms of the *specific radioactivity* (SR) of the plasma creatinine, defined as the radioactivity per millimole. The data taken, then, will be a

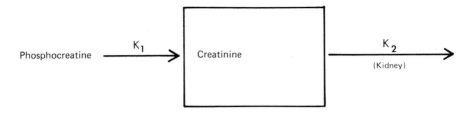

Phosphocreatine $\xrightarrow{K_1}$ | Creatinine | $\xrightarrow{K_2}$ (Kidney)

Fig. 15.4 Model of creatinine formation and excretion, with phosphocreatine as the precursor and the creatinine simply eliminated by the kidney. A single compartment with a single entry (k_1) and a single exit (k_2).

series of samples at varying times after administration of a single dose of labeled creatinine, each sample assayed for total creatinine concentration and radioactivity. The data obtained will conform to the equation

$$SR = SR_0 e^{-kt} \qquad \text{(Eq. 15.10)}$$

and the shape of the SR vs. time plot will be like the examples of Figs. 15.2 and 15.3. Extrapolation of the regression fitted to the semi-log plot to time zero will yield the SR at time zero, and

$$\text{Dose}/SR_0 = \text{total pool size} \qquad \text{(Eq. 15.11)}$$

Assuming a steady state, the rate of creatinine formation is equal to the rate of excretion, both of which are given by the product of k and the initial creatinine concentration.

15.4. TWO-COMPARTMENT KINETICS

15.4.1. With Single Loss from Compartment 2

When two compartments exchange material in a dynamic steady state, and when there is a one-way loss of material from the second compartment, the method of analysis of tracer data follows the same general ideas given above. As a model, the system in Fig. 15.5 is a compartment model of glucose breakdown via pyruvate to CO_2. This model is admittedly naive but will serve to illustrate the ideas. The glucose compartment is assumed to have a single entry flow equal to the net outflow. In the following paragraphs, a step-by-step analysis of the information gained from a single injection of ^{14}C-labeled glucose will be outlined.

Let us assume that we inject a dose of ^{14}C-labeled glucose equal to 25,000 dpm radioactivity, and that this injected dose becomes well mixed in the

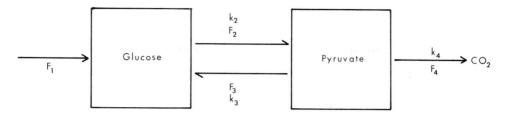

Fig. 15.5 Diagram of a two-compartment model for glucose metabolism first to pyruvate and then to CO_2. In this initial diagram, only the pathways are shown, with the rate constants (k) and flows (F) labelled.

glucose compartment within an acceptably short time. Over the next 30 min a series of samples are obtained, each of which is assayed for glucose concentration and ^{14}C radioactivity. The data are expressed as SR, i.e., the radioactivity in dpm per milligram of carbon contained in glucose. The data which might be obtained for this hypothetical exercise are shown in Fig. 15.6, plotted in semi-log form. Unlike the simple exponential examples studied previously, the data do not form a straight line, at least for the initial time period. By a simple graphic method called *curve peeling* or *stripping*, the data may be broken down into two linear components. The first step in this process is to observe that the final portion of the data, say from 15 to 25 min in Fig. 15.6, describes a straight line. The line may either be fitted by eye or calculated by simple linear regression procedures. By extrapolating the line, we obtain the zero-time intercept, which in this example is 703. Now in the early portions of the curve, the extrapolated line is subtracted, point by point, from the curve fitted through the data points. For example, at 3 min the extrapolated line gives a value of 556 and the actual data point is 922, so the difference, 366, is plotted at that time. Continuing this process produces a second straight line, with a zero-time intercept of 1798 and a slope as shown of -0.53. The complete equation for the curvilinear data is now given by the general equation

$$SR(t) = Ae^{-at} + Be^{-bt} \qquad \text{(Eq. 15.12)}$$

which in this example gives

$$SR(t) = 1798e^{-0.53t} + 703e^{-0.076t} \qquad \text{(Eq. 15.13)}$$

Assuming we could sample the second compartment at the same times, we would observe data such as those given in Fig. 15.7. Again, the data are not initially linear, but have a linear portion later on. By estimating the late portion's slope and intercept in exactly the same fashion as for Fig. 15.6,

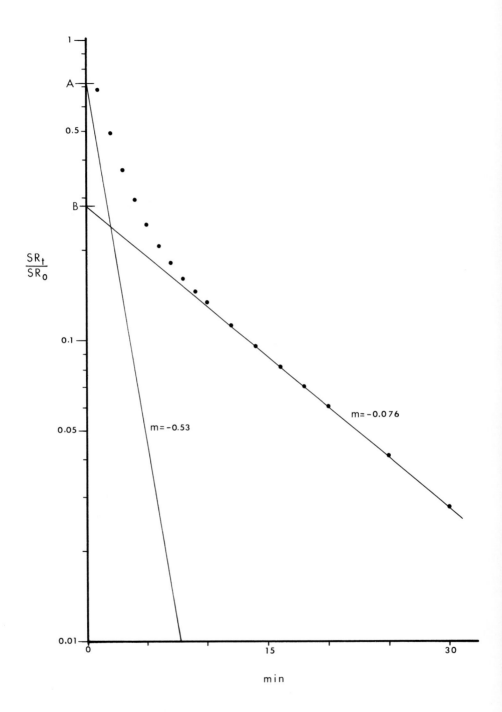

Fig. 15.6 The data for the glucose compartment in Fig. 15.5. The data points are plotted, the peeled lines shown, and the slopes and intercepts marked numerically on the graph. Data correspond to the text example.

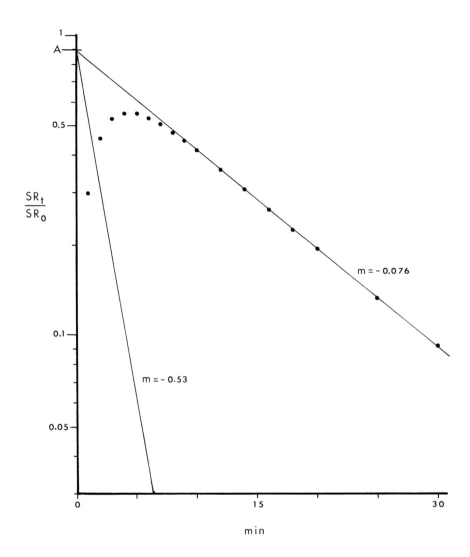

Fig. 15.7 Plot as in Fig. 15.6, but for the second compartment, pyruvate. The data are shown with lines fitted by eye.

the values shown are derived. The peeling process is only a little different, since the difference has the opposite sign but is plotted in the same way, giving the steeper second line shown. Notice that the intercept for both lines is the same, and that the slopes for the "fast" and "slow" components are the same as the slopes derived for the first compartment. This is a characteristic of two-compartment models with no external exchanges and serves as a test of this sort of data. The equation for the time course of SR in the second compartment is given by

$$SR(t) = -2235e^{-0.53t} + 2235e^{-0.076t} \qquad \text{(Eq. 15.14)}$$

Calculation of the Pool Size.   The pool size for the first compartment may be calculated simply as the total dose of radioactivity divided by the sum of the zero-time intercepts:

$$G = D/(A + B) \qquad \text{(Eq. 15.15)}$$

where G is the glucose pool in mg C, A and B are the terms from Eq. (15.13), and D is the total dose. For our example, since D was 25,000, the pool is:

$$G = 25,000/(703 + 1798) = 10 \text{ g} \qquad \text{(Eq. 15.16)}$$

The pool size for pyruvate is more complicated and is given later on.

Calculation of Rates and Flows.   Although the exponents and intercept terms of Eq. (15.13) do not correspond directly to any rate or flow shown in Fig. 15.5, they can be used to derive them. From a more formal derivation of the equations describing the system (Shipley & Clark, 1972), we have

$$k_2 = (aA + bB)/(A + B) = 0.4 \qquad \text{(Eq. 15.17)}$$

using values given above for a, b, A, and B. Also, the sum of k_3 and k_4 is given by

$$k_3 + k_4 = a + b - k_2 = 0.2 \qquad \text{(Eq. 15.18)}$$

again using the values given above. This sum may be used in the equation for k_3:

$$k_3 = \frac{k_2(k_3 + k_4) - ab}{k_2} = \frac{(0.4)(0.2) - (0.53)(0.076)}{0.4} = 0.1$$
$$\text{(Eq. 15.19)}$$

and by substituting this value for k_3 into Eq. (15.18), the value for k_4 is also seen to be 0.1. These calculations completely characterize the rate constants for the model of Fig. 15.5 but do not give the actual material flow rates.

Flow Rates.   From the glucose pool size and the value for k_2, we may calculate the first flow:

$$F_2 = k_2G = (10)(0.4) = 4 \text{ g C/hr} \qquad \text{(Eq. 15.20)}$$

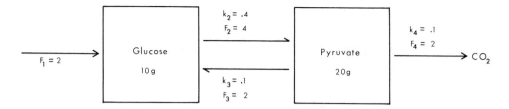

Fig. 15.8 The final model calculations, derived from the data in Figs. 15.6 and 15.7, as explained in the text.

Now by writing balance equations for the pyruvate compartment

$$F_2 - k_3 P = k_4 P \qquad \text{(Eq. 15.21)}$$

and so

$$4 - (0.1)P = (0.1)P; \quad P = 20 \text{ g} \qquad \text{(Eq. 15.22)}$$

giving us the pyruvate pool size (P) in grams of carbon. The last flow rates are given by

$$F_3 = k_3 P = (0.1)(20) = 2 \text{ g C/hr} \qquad \text{(Eq. 15.23)}$$

and

$$F_4 = k_4 P = (0.1)(20) = 2 \text{ g C/hr} \qquad \text{(Eq. 15.24)}$$

Since $F_1 = F_4$, the compartment model is complete and is shown again in Fig. 15.8 with all compartment sizes, rate constants, and material flow rates as calculated in the earlier example.

The foregoing example serves to illustrate the power of the tracer technique. By injecting a single known dose of tracer, the pool sizes, rate constants, and material flows for each pathway were calculated. Other statistics, such as the turnover rate, may also be derived from the data. A few points should probably be emphasized, however, to caution against possible errors. In order to perform the curve stripping and slope/intercept calculations, the data set must be fairly extensive and the points close together in time, especially during the early portion of the experiment. Small errors on a log scale can combine through a series of calculations to produce large errors in the final results. It is also important that the pathways be known. The presence of another exchange pathway in the model just given would significantly alter the kinetic curve obtained. Conversely, use of the analysis scheme above when the model assumptions are not true will yield spurious results.

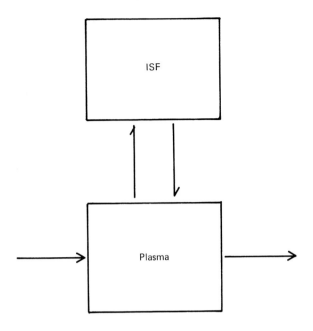

Fig. 15.9 A two-compartment model for the extracellular space in a vertebrate, showing the plasma and interstitial fluids (ISF). Entry and loss are from the same compartment, with a "side" compartment exchanging.

15.4.2. Inulin as an Extracellular Marker

A common tracer application in physiology is the use of various large molecules like inulin to estimate the size of the extracellular fluid (ECF) pool and to follow its turnover by urine formation. In this case, for a typical vertebrate, we may model the fluid compartments as shown in Fig. 15.9. As in the previous example, there are two compartments, usually corresponding to the plasma volume and the interstitial fluid compartment, but the fluid entry and loss both occur in the plasma compartment, the interstitial fluid compartment being dead-ended.

Inulin is a carbohydrate molecule of about 5000 MW and is usually supplied in frozen crystalline form. It has a strong tendency to hydrolyze spontaneously in solution, and so should be used within a short time after thawing and dissolution. The reason for this is that the breakdown fragments may be sufficiently small as to either enter the intracellular compartment or be metabolized, causing an error in the observed distribution in either case. An alternative to the frequent purchase of new material is to purify it on a chromatographic column.

The injection of a single dose of labeled inulin will usually produce an exponential curve similar to that of Fig. 15.3, but often there is an initial non-linear portion which represents not exchange with another pool, but mixing time within the extracellular fluid space. Extrapolation of the linear portion (on a semi-log plot) provides an estimate of the zero-time concentration, and division of the total dose by the zero-time concentration provides an estimate of the *inulin space*. The rate of disappearance of the inulin, i.e., the slope of the semi-log plot, provides a rate constant for clearance, as in the analyses above. Provided that it is determined that urine formation is the only route of loss, the inulin clearance is equal to the urine filtration rate. In some circumstances, however, inulin has been reported to cross the gills of aquatic animals (Kirschner, 1980); when this is true, the clearance will over-estimate the rate of urine formation. Whether the gills of fish retain inulin or not may be a function of stress and experimental conditions; in studies with catheterized fish, the gill loss has been negligible (Cameron and Wood, 1978; Cameron, 1980).

Several other substances have been used as extracellular markers in the same fashion as inulin, including raffinose, sucrose, mannitol, and thiosulfate. The distribution space estimates obtained with each of these is somewhat different, roughly in proportion to their molecular weight, with the smallest compounds yielding the largest estimates. Since the larger compounds diffuse more slowly, they may underestimate the volume of slowly exchanging regions of the ECF, such as connective tissue. On the other hand, any permeability of the cells to the smaller compounds will lead to over-estimates of the ECF. Thus it is probably more accurate to report data as inulin space, sucrose space, etc., rather than ECF volume when such techniques are employed.

15.5. THE USE OF TRACERS FOR ION MOVEMENTS

The preceding discussion focused on the use of labeled organic compounds for determination of pathways, compartment volumes, rate constants, etc. Another extensive area of application of radioisotope techniques, however, is in the study of ion movements, employing radioisotopes of various common salts (see Table 14.1). To take just one example, the rate of active uptake of sodium by freshwater animals is a typical application.

In the example illustrated in Fig. 15.10, an animal of known weight is enclosed in a measured water volume containing Na^+. At time zero, a carefully measured dose (D) of $^{22}Na^+$ is added to the water, and for some period of time afterward samples are withdrawn for analysis of sodium concentration, $[Na^+]$, and radioactivity, $^*Q$. The data are expressed as specific radioactivity:

$$SR = {}^*Q/[Na^+] \qquad \text{(Eq. 15.25)}$$

After the initial mixing period in the water, the data will fit a simple first-order exponential, as shown in Fig. 15.10, except that after a while the data will become non-linear on a semi-log plot. The reason for this is that the movement of Na^+ is two-way, both entering the animal by (active) uptake mechanisms and leaving it by other pathways. At first the radioactivity is all outside, but as the experiment progresses, the specific activity of the internal sodium pool increases and back-flux becomes significant, leading to the non-linearity of the observed data.

A mathematical description of this system is possible without requiring a steady state, i.e., when there is a net uptake or loss of Na^+ by the animal. If the flux from the outside to the inside is denoted by J_i, the flux from the inside to the outside as J_o, and the net flux as J_{net}, then the rate of change of radioactivity is described by

$$d^*Q_o/dt = J_o SR_o - J_i SR_i \qquad \text{(Eq. 15.26)}$$

Since the net movement is a function of the difference between influx and efflux

$$J_{net} = J_i - J_o \qquad \text{(Eq. 15.27)}$$

and J_o can be eliminated from Eq. (15.26) by substitution:

$$J_i = -[(d^*Q_o/dt) - J_{net}SR_i]/(SR_o - SR_i) \qquad \text{(Eq. 15.28)}$$

This equation is quite general, not depending on the steady state, number of compartments, or whether the isotope is introduced externally or internally. The SR_i must be determined by blood sampling, but if the experiment is limited to a sufficiently short time, the internal SR remains low enough so that it can be ignored. Equation (15.28) then simplifies to

$$J_i = -(d^*Q_o/dt)/SR_o \qquad \text{(Eq. 15.29)}$$

The formal solution of these equations by integration can get rather complex, but the information needed can be rather simply obtained by graphical or arithmetic examination of the data (cf. Kirschner, 1970).

The equations above may also be applied without modification for the study of ion transport in isolated tissues or organs and for any specific ion for which a useful isotope is readily available. In sea water somewhat different techniques are required, however, simply because of the large quantity of salt contained in the medium. Even with high flux rates, the change in SR in the medium is so small that the slopes cannot be estimated with any

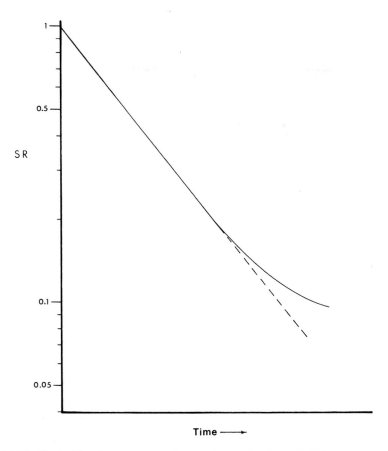

Fig. 15.10 Typical data from an ion uptake experiment, showing an initial linear portion and a later portion curving up from backflux effects. Semi-log plot.

confidence, and net flux is similarly difficult to measure against a large background salt concentration. Generally, the only satisfactory way to follow flux rates in seawater systems is by serial sampling of the blood.

15.6. MULTI-COMPARTMENT ANALYSES

Although the principles involved in the use of radioisotope tracers in multi-compartmental analysis are exactly the same as those outlined for one- and two-compartment systems above, the mathematics can become formidably complicated, and the quality and quantity of data necessary for confidence in the analysis also increase. For applications of these principles, the reader is referred to one of the specialized treatises on compartmental kinetic analysis.

LITERATURE CITED

Cameron, J. N. 1980. Body fluid pools, kidney function, and acid–base regulation in the freshwater catfish *Ictalurus punctatus*. J. Exp. Biol. 86: 171–185.

Cameron, J. N., & C. M. Wood. 1978. Renal function and acid–base regulation in two Amazonian erythrinid fishes: *Hoplias malabaricus*, a water breather, and *Hoplerythrinus unitaeniatus*, a facultative air breather. Can. J. Zool. 56: 917–930.

Kirschner, L. B. 1970. The study of NaCl transport in aquatic animals. Amer. Zool. 10: 365–376.

Kirschner, L. B. 1980. Use and limitations of inulin and mannitol for monitoring gill permeability changes in the rainbow trout. J. Exp. Biol. 85: 203–211.

Shipley, R. A., & R. E. Clark. 1972. Tracer Methods for *in Vivo* Kinetics. Academic Press, New York. 239 pp.

SUGGESTED FURTHER READING

Milhorn, H. T., Jr. 1966. The application of control theory to physiological systems. W. B. Saunders Co., Philadelphia. 386 pp.

APPENDICES

APPENDIX 1

UNITS OF PRESSURE AND CONVERSION FACTORS[a]

	atm[b]	bar[c]	psi[d]	torr[e]	mm H_2O	Pa[f]
1 atm		1.013	14.70	760.	10333.	$1.013 \cdot 10^5$
1 bar	$9.869 \cdot 10^{-1}$		14.50	750.1	10197.	$1.000 \cdot 10^5$
1 psi	$6.805 \cdot 10^{-2}$	$6.895 \cdot 10^{-2}$		51.71	703.1	$6.895 \cdot 10^3$
1 torr	$1.316 \cdot 10^{-3}$	$1.333 \cdot 10^{-3}$	$1.934 \cdot 10^{-2}$		13.60	$1.333 \cdot 10^2$
1 mm H_2O	$9.678 \cdot 10^{-5}$	$9.806 \cdot 10^{-5}$	$1.422 \cdot 10^{-3}$	$7.355 \cdot 10^{-2}$		9.806
1 Pa	$9.869 \cdot 10^{-6}$	$1.000 \cdot 10^{-5}$	$1.450 \cdot 10^{-4}$	$7.501 \cdot 10^{-3}$	0.1020	

[a]Modified, with permission, from Dejours, P. 1975. Principles of Comparative Respiratory Physiology. Elsevier: Amsterdam. 253 pp.
[b]atm = standard atmosphere.
[c]bar = 10^6 baryes = 10^6 dynes·cm^{-2}.
[d]psi = pounds·in^{-2}.
[e]torr = mm Hg.
[f]Pa = Pascal = N·m^{-2}.

APPENDIX 2

SOLUBILITY OF O_2 AND CO_2 AT DIFFERENT TEMPERATURES AND SALINITES

APPENDIX TABLE 2.1
The Volume Solubility of Oxygen at a Standard Atmosphere as a Function of Salinity (in Parts per Thousand) and Temperature in °C[a,b]

Temp:					Salinity					
	0	5	10	15	20	25	30	35	40	45
0	10.35	9.99	9.64	9.30	8.98	8.67	8.36	8.07	7.79	7.52
1	10.07	9.72	9.38	9.06	8.75	8.45	8.16	7.87	7.60	7.34
2	9.79	9.46	9.14	8.83	8.53	8.24	7.96	7.68	7.42	7.17
3	9.53	9.21	8.90	8.60	8.31	8.03	7.76	7.50	7.25	7.01
4	9.28	8.98	8.68	8.39	8.11	7.84	7.58	7.33	7.08	6.85
5	9.05	8.75	8.46	8.18	7.91	7.65	7.40	7.16	6.92	6.69
6	8.82	8.53	8.25	7.98	7.72	7.47	7.23	6.99	6.77	6.55
7	8.60	8.32	8.05	7.79	7.54	7.30	7.06	6.84	6.62	6.40
8	8.38	8.12	7.86	7.61	7.37	7.13	6.90	6.68	6.47	6.27
9	8.18	7.92	7.67	7.43	7.20	6.97	6.75	6.54	6.33	6.13
10	7.99	7.74	7.50	7.26	7.03	6.81	6.60	6.40	6.20	6.00
11	7.80	7.56	7.32	7.10	6.88	6.66	6.46	6.26	6.06	5.88
12	7.62	7.39	7.16	6.94	6.73	6.52	6.32	6.13	5.94	5.76
13	7.45	7.22	7.00	6.79	6.58	6.38	6.19	6.00	5.82	5.64
14	7.28	7.06	6.85	6.64	6.44	6.25	6.06	5.87	5.70	5.53
15	7.12	6.91	6.70	6.50	6.30	6.12	5.93	5.75	5.58	5.42
16	6.96	6.76	6.56	6.36	6.17	5.99	5.81	5.64	5.47	5.31
17	6.81	6.61	6.42	6.23	6.05	5.87	5.69	5.53	5.36	5.21
18	6.67	6.48	6.29	6.10	5.92	5.75	5.58	5.42	5.26	5.10

(Continued)

APPENDIX TABLE 2.1 (Continued)
The Volume Solubility of Oxygen at a Standard Atmosphere as a Function of Salinity (in Parts per Thousand) and Temperature in °C[a,b]

					Salinity					
Temp:	0	5	10	15	20	25	30	35	40	45
19	6.53	6.34	6.16	5.98	5.80	5.64	5.47	5.31	5.16	5.01
20	6.40	6.21	6.03	5.86	5.69	5.53	5.37	5.21	5.06	4.91
21	6.27	6.09	5.91	5.74	5.58	5.42	5.26	5.11	4.96	4.82
22	6.15	5.97	5.80	5.63	5.47	5.31	5.16	5.01	4.87	4.73
23	6.03	5.85	5.69	5.52	5.37	5.21	5.07	4.92	4.78	4.64
24	5.91	5.74	5.58	5.42	5.27	5.12	4.97	4.83	4.69	4.56
25	5.80	5.63	5.47	5.32	5.17	5.02	4.88	4.74	4.61	4.48
26	5.69	5.53	5.37	5.22	5.07	4.93	4.79	4.66	4.52	4.40
27	5.58	5.43	5.27	5.12	4.98	4.84	4.70	4.57	4.44	4.32
28	5.48	5.33	5.18	5.03	4.89	4.75	4.62	4.49	4.37	4.24
29	5.38	5.23	5.08	4.94	4.80	4.67	4.54	4.41	4.29	4.17
30	5.29	5.14	4.99	4.85	4.72	4.59	4.46	4.34	4.21	4.10
31	5.19	5.05	4.91	4.77	4.64	4.51	4.38	4.26	4.14	4.03
32	5.10	4.96	4.82	4.69	4.56	4.43	4.31	4.19	4.07	3.96
33	5.01	4.87	4.74	4.61	4.48	4.35	4.23	4.12	4.00	3.89
34	4.93	4.79	4.66	4.53	4.40	4.28	4.16	4.05	3.93	3.82
35	4.85	4.71	4.58	4.45	4.33	4.21	4.09	3.98	3.87	3.76
36	4.76	4.63	4.50	4.38	4.26	4.14	4.02	3.91	3.80	3.70
37	4.69	4.56	4.43	4.30	4.19	4.07	3.96	3.84	3.74	3.63
38	4.61	4.48	4.36	4.23	4.12	4.00	3.89	3.78	3.68	3.57
39	4.53	4.41	4.28	4.16	4.05	3.94	3.83	3.72	3.61	3.51
40	4.46	4.34	4.21	4.10	3.98	3.87	3.76	3.66	3.55	3.45

[a]Data derived from equations of Green, E. J., and D. E. Carritt. 1967. New tables for oxygen saturation of water. J. Mar. Res. 25:140–147.
[b]Solubility is expressed as ml O_2 (STPD) per liter of water equilibrated with a water-saturated atmosphere of 20.94% O_2.

APPENDIX TABLE 2.2
The Volume Solubility of Oxygen per Partial Pressure Unit as a Function of Salinity (in Parts per Thousand) and Temperature in °C[a,b]

					Salinity					
Temp:	0	5	10	15	20	25	30	35	40	45
0	7.00	6.75	6.52	6.29	6.07	5.86	5.65	5.46	5.27	5.08
1	6.80	6.57	6.34	6.12	5.91	5.71	5.51	5.32	5.14	4.96
2	6.62	6.40	6.18	5.97	5.76	5.57	5.38	5.19	5.02	4.85

(Continued)

APPENDIX TABLE 2.2 (Continued)
The Volume Solubility of Oxygen per Partial Pressure Unit as a Function of Salinity (in Parts per Thousand) and Temperature in °C[a,b]

Temp:	Salinity									
	0	5	10	15	20	25	30	35	40	45
3	6.44	6.23	6.02	5.82	5.62	5.43	5.25	5.07	4.90	4.74
4	6.28	6.07	5.87	5.67	5.48	5.30	5.12	4.95	4.79	4.63
5	6.11	5.91	5.72	5.53	5.35	5.17	5.00	4.84	4.68	4.53
6	5.96	5.77	5.58	5.40	5.22	5.05	4.89	4.73	4.57	4.43
7	5.81	5.62	5.44	5.27	5.10	4.93	4.77	4.62	4.47	4.33
8	5.67	5.49	5.31	5.14	4.98	4.82	4.67	4.52	4.37	4.24
9	5.53	5.36	5.19	5.02	4.87	4.71	4.56	4.42	4.28	4.14
10	5.40	5.23	5.07	4.91	4.76	4.61	4.46	4.32	4.19	4.06
11	5.27	5.11	4.95	4.80	4.65	4.51	4.37	4.23	4.10	3.97
12	5.15	4.99	4.84	4.69	4.55	4.41	4.27	4.14	4.01	3.89
13	5.03	4.88	4.73	4.59	4.45	4.31	4.18	4.05	3.93	3.81
14	4.92	4.77	4.63	4.49	4.35	4.22	4.09	3.97	3.85	3.74
15	4.81	4.67	4.53	4.39	4.26	4.13	4.01	3.89	3.77	3.66
16	4.71	4.57	4.43	4.30	4.17	4.05	3.93	3.81	3.70	3.59
17	4.61	4.47	4.34	4.21	4.09	3.97	3.85	3.74	3.63	3.52
18	4.51	4.38	4.25	4.12	4.00	3.89	3.77	3.66	3.55	3.45
19	4.42	4.29	4.16	4.04	3.92	3.81	3.70	3.59	3.49	3.38
20	4.33	4.20	4.08	3.96	3.85	3.73	3.63	3.52	3.42	3.32
21	4.24	4.12	4.00	3.88	3.77	3.66	3.56	3.45	3.36	3.26
22	4.15	4.04	3.92	3.81	3.70	3.59	3.49	3.39	3.29	3.20
23	4.07	3.96	3.84	3.73	3.63	3.52	3.42	3.33	3.23	3.14
24	3.99	3.88	3.77	3.66	3.56	3.46	3.36	3.27	3.17	3.08
25	3.92	3.81	3.70	3.60	3.49	3.39	3.30	3.21	3.11	3.03
26	3.84	3.74	3.63	3.53	3.43	3.33	3.24	3.15	3.06	2.97
27	3.77	3.67	3.56	3.46	3.37	3.27	3.18	3.09	3.00	2.92
28	3.70	3.60	3.50	3.40	3.31	3.21	3.12	3.04	2.95	2.87
29	3.64	3.54	3.44	3.34	3.25	3.16	3.07	2.98	2.90	2.82
30	3.57	3.47	3.38	3.28	3.19	3.10	3.01	2.93	2.85	2.77
31	3.51	3.41	3.32	3.22	3.13	3.05	2.96	2.88	2.80	2.72
32	3.45	3.35	3.26	3.17	3.08	2.99	2.91	2.83	2.75	2.67
33	3.39	3.29	3.20	3.11	3.03	2.94	2.86	2.78	2.70	2.63
34	3.33	3.24	3.15	3.06	2.98	2.89	2.81	2.73	2.66	2.58
35	3.28	3.18	3.10	3.01	2.93	2.84	2.77	2.69	2.61	2.54
36	3.22	3.13	3.04	2.96	2.88	2.80	2.72	2.64	2.57	2.50
37	3.17	3.08	2.99	2.91	2.83	2.75	2.67	2.60	2.53	2.46
38	3.12	3.03	2.94	2.86	2.78	2.70	2.63	2.56	2.48	2.42
39	3.06	2.98	2.90	2.82	2.74	2.66	2.59	2.51	2.44	2.37
40	3.02	2.93	2.85	2.77	2.69	2.62	2.54	2.47	2.40	2.34

[a]Data derived from equations of Green, E. J., and D. E. Carritt. 1967. New tables for oxygen saturation of seawater. J. Mar. Res. 25: 140–147.
[b]Solubility is expressed as milliliters of O_2 (STPD) $\times 10^5$ per milliliter of water per torr O_2 partial pressure.

APPENDIX TABLE 2.3
The Molar Solubility of Oxygen per Partial Pressure Unit as a Function of Salinity (in Parts per Thousand) and Temperature in °C[a,b]

Temp:	Salinity									
	0	5	10	15	20	25	30	35	40	45
0	3.12	3.01	2.91	2.81	2.71	2.61	2.52	2.44	2.35	2.27
1	3.04	2.93	2.83	2.73	2.64	2.55	2.46	2.38	2.29	2.22
2	2.96	2.86	2.76	2.66	2.57	2.49	2.40	2.32	2.24	2.16
3	2.88	2.78	2.69	2.60	2.51	2.42	2.34	2.26	2.19	2.11
4	2.80	2.71	2.62	2.53	2.45	2.37	2.29	2.21	2.14	2.07
5	2.73	2.64	2.55	2.47	2.39	2.31	2.23	2.16	2.09	2.02
6	2.66	2.57	2.49	2.41	2.33	2.25	2.18	2.11	2.04	1.98
7	2.59	2.51	2.43	2.35	2.28	2.20	2.13	2.06	2.00	1.93
8	2.53	2.45	2.37	2.30	2.22	2.15	2.08	2.02	1.95	1.89
9	2.47	2.39	2.32	2.24	2.17	2.10	2.04	1.97	1.91	1.85
10	2.41	2.33	2.26	2.19	2.12	2.06	1.99	1.93	1.87	1.81
11	2.35	2.28	2.21	2.14	2.08	2.01	1.95	1.89	1.83	1.77
12	2.30	2.23	2.16	2.09	2.03	1.97	1.91	1.85	1.79	1.74
13	2.25	2.18	2.11	2.05	1.99	1.93	1.87	1.81	1.76	1.70
14	2.20	2.13	2.07	2.00	1.94	1.88	1.83	1.77	1.72	1.67
15	2.15	2.08	2.02	1.96	1.90	1.85	1.79	1.74	1.68	1.63
16	2.10	2.04	1.98	1.92	1.86	1.81	1.75	1.70	1.65	1.60
17	2.06	2.00	1.94	1.88	1.82	1.77	1.72	1.67	1.62	1.57
18	2.01	1.95	1.90	1.84	1.79	1.74	1.68	1.63	1.59	1.54
19	1.97	1.91	1.86	1.80	1.75	1.70	1.65	1.60	1.56	1.51
20	1.93	1.88	1.82	1.77	1.72	1.67	1.62	1.57	1.53	1.48
21	1.89	1.84	1.78	1.73	1.68	1.64	1.59	1.54	1.50	1.45
22	1.85	1.80	1.75	1.70	1.65	1.60	1.56	1.51	1.47	1.43
23	1.82	1.77	1.72	1.67	1.62	1.57	1.53	1.49	1.44	1.40
24	1.78	1.73	1.68	1.64	1.59	1.54	1.50	1.46	1.42	1.38
25	1.75	1.70	1.65	1.60	1.56	1.52	1.47	1.43	1.39	1.35
26	1.72	1.67	1.62	1.58	1.53	1.49	1.45	1.41	1.37	1.33
27	1.68	1.64	1.59	1.55	1.50	1.46	1.42	1.38	1.34	1.30
28	1.65	1.61	1.56	1.52	1.48	1.43	1.39	1.36	1.32	1.28
29	1.62	1.58	1.53	1.49	1.45	1.41	1.37	1.33	1.29	1.26
30	1.59	1.55	1.51	1.47	1.42	1.38	1.35	1.31	1.27	1.24
31	1.57	1.52	1.48	1.44	1.40	1.36	1.32	1.29	1.25	1.21
32	1.54	1.50	1.46	1.41	1.38	1.34	1.30	1.26	1.23	1.19
33	1.51	1.47	1.43	1.39	1.35	1.31	1.28	1.24	1.21	1.17
34	1.49	1.45	1.41	1.37	1.33	1.29	1.26	1.22	1.19	1.15
35	1.46	1.42	1.38	1.34	1.31	1.27	1.23	1.20	1.17	1.13
36	1.44	1.40	1.36	1.32	1.28	1.25	1.21	1.18	1.15	1.12
37	1.41	1.37	1.34	1.30	1.26	1.23	1.19	1.16	1.13	1.10
38	1.39	1.35	1.31	1.28	1.24	1.21	1.17	1.14	1.11	1.08
39	1.37	1.33	1.29	1.26	1.22	1.19	1.15	1.12	1.09	1.06
40	1.35	1.31	1.27	1.24	1.20	1.17	1.14	1.10	1.07	1.04

[a]Data derived from equations of Green, E. J., and D. E. Carritt. 1967. New tables for oxygen saturation of seawater. J. Mar. Res. 25: 140–147.

[b]Solubility is expressed as micromoles of O_2 per milliliter of water per torr O_2 partial pressure.

APPENDIX TABLE 2.4

The Solubility of Carbon Dioxide as a Function of Temperature and Salinity[a]

Temp:	Salinity					
	0	10	20	30	35	40
0	7.758	7.364	6.990	6.635	6.465	6.298
2	7.174	6.813	6.469	6.143	5.986	5.833
4	6.650	6.317	6.001	5.701	5.557	5.416
6	6.178	5.871	5.580	5.303	5.170	5.040
8	5.751	5.469	5.200	4.945	4.822	4.702
10	5.366	5.105	4.857	4.621	4.507	4.396
12	5.017	4.776	4.546	4.327	4.222	4.119
14	4.700	4.477	4.264	4.062	3.964	3.869
16	4.412	4.205	4.008	3.820	3.729	3.641
18	4.149	3.958	3.775	3.600	3.516	3.434
20	3.910	3.732	3.562	3.400	3.322	3.245
22	3.691	3.526	3.368	3.217	3.144	3.073
24	3.491	3.337	3.190	3.050	2.982	2.915
26	3.307	3.164	3.027	2.897	2.833	2.771
28	3.138	3.005	2.878	2.756	2.697	2.639
30	2.983	2.859	2.741	2.627	2.572	2.518
32	2.840	2.725	2.615	2.509	2.457	2.407
34	2.708	2.601	2.498	2.400	2.352	2.305
36	2.587	2.487	2.391	2.299	2.254	2.211
38	2.474	2.382	2.292	2.207	2.165	2.124
40	2.370	2.284	2.201	2.121	2.082	2.044

[a]Units are 10^2 mol L^{-1} atm^{-1}, salinity in parts per thousand, and temperature in °C.

APPENDIX TABLE 2.5

The Solubility of Carbon Dioxide as a Function of Temperature and Salinity[a]

Temp:	Salinity					
	0	10	20	30	35	40
0	102.08	96.89	91.97	87.30	85.07	82.87
2	94.39	89.64	85.12	80.83	78.76	76.75
4	87.50	83.12	78.96	75.01	73.12	71.26
6	81.29	77.25	73.42	69.78	68.03	66.32
10	70.61	67.17	63.91	60.80	59.30	57.84
12	66.01	62.84	59.82	56.93	55.55	54.20
14	61.84	58.91	56.11	53.45	52.16	50.91
16	58.05	55.33	52.74	50.26	49.07	47.91

(Continued)

APPENDIX TABLE 2.5 (Continued)
The Solubility of Carbon Dioxide as a Function of Temperature and Salinity[a]

Temp:	Salinity					
	0	10	20	30	35	40
18	54.59	52.08	49.67	47.37	46.26	45.18
20	51.45	49.11	46.87	44.74	43.71	42.70
22	48.57	46.39	44.32	42.33	41.37	40.43
24	45.93	43.91	41.97	40.13	39.24	38.36
26	43.51	41.63	39.83	38.12	37.28	36.46
28	41.29	39.54	37.87	36.26	35.49	34.72
30	39.25	37.62	36.07	34.57	33.84	33.13
32	37.37	35.86	34.41	33.01	32.33	31.67
34	35.63	34.22	32.87	31.58	30.95	30.33
36	34.04	32.72	31.46	30.25	29.66	29.09
38	32.55	31.34	30.16	29.04	28.49	27.95
40	31.18	30.05	28.96	27.91	27.39	26.89

[a]Units are μmol L^{-1} torr^{-1}, salinity in parts per thousand, and temperature in °C.

APPENDIX 3

VAPOR PRESSURE OF WATER AT DIFFERENT TEMPERATURES

APPENDIX TABLE 3.1
The Vapor Pressure of Water (P_w) in a Water-Saturated Atmosphere at Various Temperatures[a,b]

Temp	P_w	Temp	P_w	Temp	P_w	Temp	P_w
0	4.6	10	9.2	20	17.5	30	31.8
1	4.9	11	9.8	21	18.7	31	33.7
2	5.3	12	10.5	22	19.8	32	35.7
3	5.7	13	11.2	23	21.1	33	37.7
4	6.1	14	12.0	24	22.4	34	39.9
5	6.5	15	12.8	25	23.8	35	42.2
6	7.0	16	13.6	26	25.2	36	44.6
7	7.5	17	14.5	27	26.7	37	47.1
8	8.0	18	15.5	28	28.3	38	49.7
9	8.6	19	16.5	29	30.0	39	52.4
10	9.2	20	17.5	30	31.8	40	5.3

[a]Data from the *Handbook of Chemistry and Physics*.
[b]The vapor pressure is given in torr and the temperature in °C.

APPENDIX 4

***BASIC* COMPUTER PROGRAM FOR
CALCULATING THE TIME COURSE OF
WASH OUT OR CHANGEOVER FROM
ONE CONCENTRATION TO ANOTHER[1]**

```
10  '*****************************************
20  '*    Washout Model to Show The        *
30  '*    Time Course of Change in         *
40  '*    Concentration. J. N. Cameron     *
50  '*              12 January 1984        *
60  '*****************************************
90  CLEAR 500
100 PRINT "This is a program designed to calculate the change:
110 PRINT "in PO2 in water within a closed respirometer of"
120 PRINT "Initial PO2 and water of New PO2 flowing through."
130 PRINT: PRINT "Enter X, Y, W, F: Volume, Init. PO2, New
        PO2, Flow"
140 INPUT X,Y,W,F
150 T = 0
160 YI = Y
170 PRINT "0                      PO2                      ";YI
180 H$ = STRING$(63,"-"): PRINT H$
```

[1]Initial and final concentrations, the system volume, and the system flow rate are entered as data, and the program simulates 90 min of time.

```
190 PRINT T;
200 PRINT TAB((62*Y)/YI);"*"
210 FOR J = 1 TO 5 STEP 1
220     Z = X*Y – (F*Y) + (F*W)
230     Y = Z/X
240 NEXT J
250 T = T + 5
260 PRINT T;
270 PRINT TAB((62*Y)/YI); "*";: PRINT USING"###.#"; Y
280 IF T = 60 GOTO 300
290 GOTO 210
300 GOTO 300
310 END
```

Sample output, with volume $= 2500$, initial $P_{O_2} = 150$, new $P_{O_2} = 50$, and flow $= 100$:

0	P_{O_2}	150
0		*
5		*131.5
10		*116.5
15		*104.2
20		* 94.2
25		* 86.0
30		* 79.4
35		* 74.0
40		* 69.5
45		* 65.9
50		* 63.0
55		* 60.6
60		* 58.6
65		* 57.0
70		* 55.7
75		* 54.7
80		* 53.8
85		* 53.1
90		* 52.5

APPENDIX 5

COMPOSITION AND CHARACTERISTICS OF SOME STANDARD BUFFERS

APPENDIX TABLE 5.1
Data for Three National Bureau of Standards Reference
Buffers at Temperatures from 0 to 40°C
Composition

Buffer	Formula	Weight, (g L^{-1})
Phthalate	$KHC_8H_4O_4$	10.12
Phosphate	KH_2PO_4	3.388
	Na_2HPO_4	3.533
Borate	$Na_2B_4O_7 \cdot 10\ H_2O$	3.80

Values for pH as a Function of Temperature

Temp. (°C)	0.05 M potassium hydrogen phthalate	0.025 M Na_2HPO_4 plus 0.025 M KH_2PO_4	0.05 M sodium tetraborate decahydrate
0	4.012	6.983	9.512
1	4.010	6.976	9.496
2	4.009	6.969	9.480
3	4.007	6.961	9.464
4	4.006	6.956	9.449

(Continued)

APPENDIX TABLE 5.1 (Continued)
Data for Three National Bureau of Standards Reference
Buffers at Temperatures from 0 to 40°C

Values for pH as a Function of Temperature

Temp. (°C)	0.05 M potassium hydrogen phthalate	0.025 M Na$_2$HPO$_4$ plus 0.025 M KH$_2$PO$_4$	0.05 M sodium tetraborate decahydrate
5	4.005	6.950	9.434
6	4.004	6.944	9.419
7	4.003	6.938	9.404
8	4.002	6.933	9.389
9	4.002	6.927	9.375
10.	4.001	6.922	9.362
11	4.001	6.916	9.350
12	4.000	6.911	9.338
13	4.000	6.906	9.326
14	4.000	6.901	9.315
15	4.000	6.896	9.305
16	4.000	6.892	9.294
17	4.000	6.888	9.282
18	4.000	6.885	9.270
19	4.001	6.881	9.259
20	4.001	6.878	9.247
21	4.002	6.874	9.236
22	4.002	6.870	9.226
23	4.003	6.867	9.216
24	4.004	6.863	9.206
25	4.005	6.860	9.196
26	4.006	6.857	9.187
27	4.007	6.855	9.178
28	4.008	6.853	9.169
29	4.010	6.851	9.161
30	4.011	6.849	9.152
31	4.012	6.847	9.143
32	4.014	6.846	9.134
33	4.015	6.844	9.126
34	4.017	6.843	9.117
35	4.019	6.842	9.109
36	4.021	6.841	9.101
37	4.023	6.840	9.093
38	4.025	6.839	9.085
39	4.027	6.838	9.077
40	4.030	6.837	9.069

EFFECTS OF DILUTION AND ADDED SALT
ON BUFFERS AND pH

Dilution or concentration of buffer solutions will alter their pH through theoretically predictable effects on either the total solution ionic strength, the specific ion activity effects, or both. For the phosphate buffer described above, the effect of diluting a stock solution consisting of 0.050 M each of KH_2PO_4 and Na_2HPO_4 with an equal volume of water is to increase the pH by 0.088 units (Bates, 1964). In general, the effect is determined by a combination of the effects of changes in water on the buffer equilibria and the effects of changes in ionic strength on the ionized and non-ionized species in the solution.

There may also be specific salt effects, which are important in two ways in pH measurement. The salts affect buffer equilibria both by changes in ionic strength and by the specific ionic effects on the equilibria. Second, there may be a specific ion effect on the glass electrode, due to imperfect selectivity of the glass for H^+ over the competing cation. The strongest ionic effect is exerted by Na^+; this effect is small at most physiological pH values but begins to be appreciable at values above 8. Since the effect is variable from one electrode to the next, it is generally good practice to standardize the electrode with a buffer of ionic strength similar to that of unknown solutions to be measured. Data for correcting buffer values for ionic strength, as well as recipes for buffers of varying ionic strength, are given by Bates (1964) and Dawson et al. (1969).

LITERATURE CITED

Bates, R. G. 1964. Determination of pH: Theory and Practice. John Wiley & Sons, New York. 435 pp.

Dawson, R. M. C., D. C. Elliott, W. H. Elliott, & K. M. Jones. 1969. Data for Biochemical Research. 2nd ed. Clarendon Press, Oxford. 536 pp.

APPENDIX 6

NOMOGRAMS FOR ASSESSMENT OF RADIATION COUNTING ERRORS

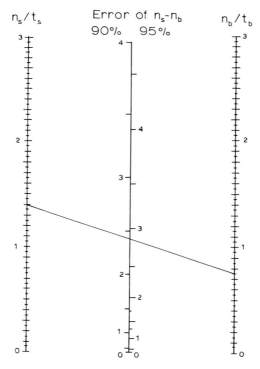

Appendix Fig. 6.1 Nomogram for the assessment of 0.9 and 0.95 error of low counting rates. To use the nomogram, first calculate n.s, the counting rate of the sample including the background, in counts per minute; t.s, the number of minutes the sample was counted; n.b, the counting rate of the background, in counts per minute; and t.b, the number of minutes the background was counted. Then draw a straight line from the point on the left-hand scale that corresponds to the quotient n.s/t.s to the point on the right-hand scale that corresponds to the quotient n.b/t.b. The point where this line crosses the center scale will correspond to the 0.9 or 0.95 error limits of the counting rate difference, i.e., n.s − n.b. A line is shown for an example in which n.s = 28, t.s = 20, n.b = 15, and t.b = 20. The line is drawn from n.s/t.s = 1.4 to n.b/t.b = 0.75, yielding the 90% error of 2.35 cpm and 95% error of 2.8 cpm. (Redrawn with permission from Wang, Y. 1969, Handbook of Radioactive Nuclides. Copyright CRC Press, Inc., Boca Raton, Florida.)

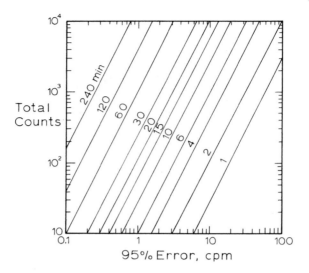

Appendix Fig. 6.2   A diagram showing the relationship between the total number of counts, time counted, and the 95% confidence level error in the total count. This diagram should be used only when the n.s/n.b ratio is large. (Redrawn with permission from Wang, Y. 1969, Handbook of Radioactive Nuclides. Copyright CRC Press, Inc., Boca Raton, Florida.)

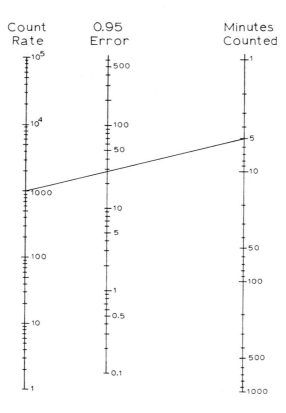

Count Rate   0.95 Error   Minutes Counted

Appendix Fig. 6.3 A nomogram for finding the 95% counting error from the total count rate (on the left) and the total minutes counted (on the right). To find the 95% error, a line is drawn from the outer scales, and the intersection value on the middle scale yields the error, as shown in the example on the drawing. (Redrawn with permission from Wang, Y. 1969, Handbook of Radioactive Nuclides. Copyright CRC Press, Inc., Boca Raton, Florida.)

APPENDIX 7

CALCULATIONS FOR DUAL-ISOTOPE COUNTING

When two radioisotopes are counted simultaneously, it will almost always be the case that the energy spectra for the two will overlap to a great or lesser degree. In order to assess the radioactivity due to each isotope independent of the other, a series of simple equations must be solved. To illustrate the procedure, we consider the case of two gamma emitters, ^{137}Cs and ^{51}Cr, whose combined spectra are shown in Fig. 14.12. Although the principal emission peaks for each isotope are fairly sharp, there is no pair of energy ranges which includes only one isotope or the other. Instead, a counter is generally set to count in two separate "windows," A and B in Fig. 4, and the following calculations performed on the data:

$$
\begin{aligned}
\text{Let A} &= \text{total counts in window A} \\
\text{B} &= \text{total counts in window B} \\
\text{Cr} &= \text{true cpm due to } ^{51}\text{Cr} \\
\text{Cs} &= \text{true cpm due to } ^{137}\text{Cs} \\
a_1 &= \text{cpm from } ^{51}\text{Cr appearing in A} \\
a_2 &= \text{cpm from } ^{51}\text{Cr appearing in B} \\
b_1 &= \text{cpm from } ^{137}\text{Cs appearing in A} \\
b2 &= \text{cpm from } ^{137}\text{Cs appearing in B}
\end{aligned}
$$

For these two windows, the total counts appearing in both can be expressed as $(a_1 + a_2)$ for Cr and the fraction appearing in A as $a_1/(a_1 + a_2) = r_1$. We also define r_2 as the fraction of counts from Cr appearing in B, r_3 as the fraction from Cs in A and r_4 as the fraction from Cs in B. Now the counts in the two windows can be expressed as

$$A = r_1\text{Cr} + r_3\text{Cs} \tag{1}$$

and

$$B = r_2 Cr + r_4 Cs \tag{2}$$

Dividing (1) by r_1 and (2) by r_2:

$$A/r_1 = Cr + (r_3/r_4)Cs \tag{3}$$

and

$$B/r_2 = Cr + (r_4/r_2)Cs \tag{4}$$

Now subtracting (4) from (3):

$$(A/r_1) - (B/r_2) = [(r_3/r_1) - (r_4/r_2)]Cs \tag{5}$$

and solving for Cs:

$$Cs = (r_2 A - r_1 B)/(r_3 r_2 - r_1 r_4) \tag{6}$$

By a similar procedure, i.e., dividing (1) by r_3 and (2) by r_4 and subtracting to solve for Cr, we arrive at the other equation required:

$$Cr = (r_4 A - r_3 B)/(r_1 r_4 - r_2 r_3) \tag{7}$$

All the A, B and r coefficients are directly calculable from counts of Cr and Cs standards individually, and may be assumed to be the same in the presence of the other isotope. Three-isotope calculations may be performed in the same fashion, setting up three equations in three unknowns and using a set of fractional coefficients determined with single-isotope standards. The calculations may be performed easily on many programmable calculators and on any small computer.

APPENDIX 8

TABLE OF ELECTRICAL SYMBOLS

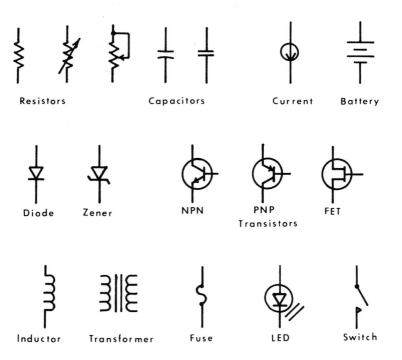

Resistors   Capacitors   Current   Battery

Diode   Zener   NPN   PNP   FET

Transistors

Inductor   Transformer   Fuse   LED   Switch

INDEX